FIBROCYTES

New Insights into
Tissue Repair and Systemic Fibroses

FIBROCYTES

New Insights into Tissue Repair and Systemic Fibroses

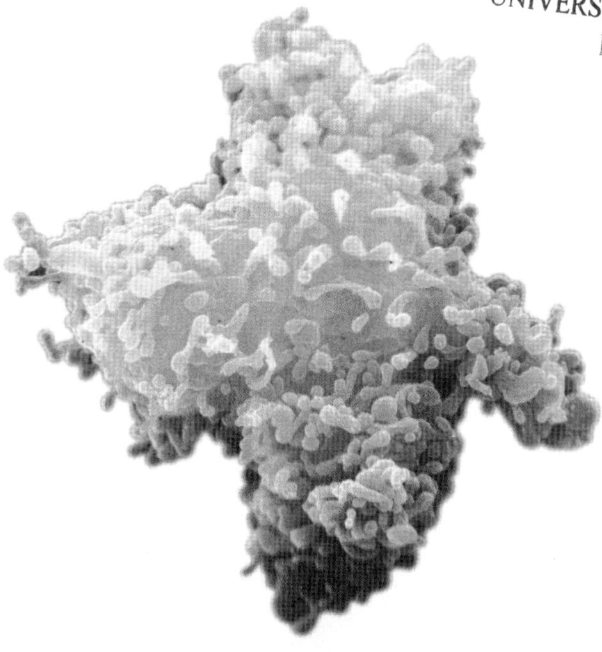

Editor

Richard Bucala

Yale University, USA

World Scientific

NEW JERSEY • LONDON • SINGAPORE • BEIJING • SHANGHAI • HONG KONG • TAIPEI • CHENNAI

Published by

World Scientific Publishing Co. Pte. Ltd.
5 Toh Tuck Link, Singapore 596224
USA office: 27 Warren Street, Suite 401-402, Hackensack, NJ 07601
UK office: 57 Shelton Street, Covent Garden, London WC2H 9HE

Library of Congress Cataloging-in-Publication Data
Fibrocytes : new insights into tissue repair and systemic fibroses / editor, Richard Bucala.
p. cm.
Includes bibliographical references and index.
ISBN-13 978-981-256-869-4 -- ISBN-10 981-256-869-7
1. Fibroblasts. 2. Fibroblasts--pathology. 3. Fibroblasts--physiology. 4. Fibroses--physiopathology. 5. Wound Healing--physiology. I. Bucala, Richard.

QP88.23 .F53 2006
612.7'5--dc22 2006051165

British Library Cataloguing-in-Publication Data
A catalogue record for this book is available from the British Library.

Typeset by Stallion Press
Email: enquiries@stallionpress.com

Printed by Mainland Press Pte Ltd

CONTRIBUTORS

Peter J. Barth, MD
Institute of Pathology
University Hospital Giessen and Marburg GmbH
Location Marburg
Medical Faculty of Philipps — University Malburg
Baldingerstraße, 35043 Malburg
Germany

Rick Bucala, MD, PhD
Yale University School of Medicine
The Anlyan Center, S525
PO Box 208031
300 Cedar Street
New Haven, CT 06520-8031
USA

Jason A. Chesney, MD, PhD
J.G. Brown Cancer Center
University of Louisville
Delia Baxter Research Bldg., Rm. 204E
508 So. Preston Street
Louisville, KY 40202
USA

Shawn E. Cowper, MD
Department of Dermatology and Pathology
Yale University School of Medicine
P.O. Box 208059
15 York St. LMP 5032
New Haven, CT 06520-8059
USA

Richard Gomer, PhD
Department of Biochemistry and Cell Biology
MS-140, Rice University
Houston, TX 77005-1892
USA

Brigitte N. Gomperts, MD
Department of Pediatrics
Division of Hematology and Oncology
University of Washington
School of Medicine in St. Louis
660 S. Euclid Avenue, Campus Box 8116
St. Louis, MO 63110
USA

Shuichi Kaneko, MD, PhD
Department of Gastroenterology and Nephrology
Kanazawa University
Graduate School of Medical Science
Takara-Machi 13-1
Kanazawa 920-8641,
Japan

Cynthia L. Kucher, MD
Greenwich Hospital — Pathology Department
5 Perryridge Road
Greenwich, CT 06830
USA

Amanda C. LaRue, PhD
Department of Veterans Affairs Medical Center
Division of Experimental Hematology
Department of Medicine
Medical University of South Carolina
Charleston, SC 29401
USA

Kouji Matsushima, MD, PhD
Department of Molecular Preventive Medicine
Graduate School of Medicine and
Faculty of Medicine
University of Tokyo
7-3-1, Hongo, Bunkyo-ku,
Tokyo, 113-0033
Japan

Sabrina Mattoli, MD, PhD
Avail Biomedical Research Institute, Basel, Switzerland
Avail GmbH
PO Box 110, CH-4003 Basel
Switzerland

Heather Medbury, PhD
Senior Scientist
Department of Surgery
University of Sydney
Westmead Hospital
Westmead 2145
Australia

Abelardo Medina
UBC Experimental Medicine
344A Jack Bell Research Centre
2660 Oak Street
Vancouver, BC
Canada

Makio Ogawa, MD, PhD
Department of Veterans Affairs Medical Center
Division of Experimental Hematology
Department of Medicine
Medical University of South Carolina
Charleston, SC 29401
USA

Darrell Pilling, PhD
Department of Biochemistry and Cell Biology
MS-140
Rice University
Houston, TX 77005-1892
USA

Arnold Postlethwaite, MD
Division of Rhematology
Department of Medicine
University of Tennessee Health Science Center
956 Court Avenue
Coleman Building, G326
Memphis, Tennessee 38163
USA

Norihiko Sakai, MD, PhD
Department of Gastroenterology and Nephrology
Kanazawa University Graduate School of Medical Science
Ishikawa
Japan

Matthhias Schmidt
Avail Biomedical Research Institute, Basel, Switzerland
Avail GmbH
PO Box 110, CH-4003 Basel
Switzerland

Paul G. Scott
Department of Surgery
2D3. 81 WMSHC
8440-112 Street
University of Alberta
Edmonton, Alberta
Canada

Robert M. Strieter, MD
Department of Internal Medicine
University of Virginia
School of Medicine
P.O. Box 800466, VA 22908-0466
USA

Edward E. Tredget, MD
Department of Surgery
2D3. 81 WMSHC
8440-112 Street
University of Alberta
Edmonton, Alberta
Canada

Takashi Wada, MD, PhD
Department of Gastroenterology and Nephrology
Kanazawa University Graduate School of Medical Science
13-1 Takara-machi
Kanazawa 920-8641
Japan

JianFei Wang
Department of Surgery
2D3.81 WMSHC
8440-112 Street
University of Alberta
Edmonton, Alberta
Canada

Yaojiong Wu
Department of Surgery
2D3. 81 WMSHC
8440-112 Street
University of Alberta
Edmonton, Alberta
Canada

CONTENTS

Chapter 1

Fibrocytes: Discovery of a Circulating Connective Tissue Cell Progenitor

Richard Bucala*

Fibrocytes are a sub-population of peripheral blood cells that produce connective tissue proteins such as collagens and α-smooth muscle actin. The identification of circulating fibrocytes has filled a void in our understanding of tissue repair, and resolved a long-standing controversy about the blood-borne origin of new fibroblasts. Fibrocytes have a prominent role in inflammatory and healing skin lesions, and in the development of tissue fibrosis. Over the last 10 years, fibrocytes have been described in granulomas, in hypertrophic scars, in pulmonary fibrosis due to asthma or acute lung injury, and in the stromal reaction to tumor invasion. Fibrocytes can further differentiate, and they are a source of the contractile myofibroblast. Fibrocytes produce abundant cytokines and contribute to tissue remodeling by secreting fibrogenic and angiogenic growth factors, and matrix metalloproteinases. The intercellular

*Professor of Medicine, and Pathology, Yale University, The School of Medicine, The Anlyan Center, S525, 300 Cedar Street, New Haven, CT 06520-8031. Tel.: 203 737 1453; Fax: 203 785 7053. E-mail: Richard.Bucala@Yale.edu

signals that modulate fibrocyte trafficking, proliferation, and differentiation are now becoming understood, and a better understanding of these signals may enable new therapies to facilitate repair and prevent pathologic tissue remodeling.

Introduction

The host response to tissue injury results ultimately in a reparative response that requires the action of connective tissue cells and their matrix products.[1] Under optimal circumstances, tissue repair produces a restoration of the normal architecture and function of the damaged tissue. In the setting of persistent inflammation, tissue invasion, or vascular or metabolic insufficiency, this reparative process may be dysregulated so that a pathologic remodeling response occurs. Remodeling may involve the replacement of normal cellular and tissue constituents by connective tissue cells and fibrosis, and it underlies the pathologic changes in lungs affected by chronic asthma or acute lung injury, chronically inflammed kidney or liver, arterial walls affected by atherosclerosis or re-stenosis injury, synovial pannus in rheumatoid arthritis, and skin affected by hypertrophic scarring.[2,3] Abnormalities in host repair also play a role in the tissue response to tumor invasion, and the stromal reaction to neoplasia has been identified as playing an important role in tumor progression.[4,5]

Until very recently, fibrosis has been considered to result from the activation, recruitment, and proliferation of locally-derived, connective tissue fibroblasts.[1,6] Mesenchymal cells are the precursors for the connective tissue cells (fibroblasts, myocytes, etc) that are the structural and supportive elements of tissue. Mesenchymal cells typically have an irregular star (stellate) or spindle (fusiform) shape with delicate branching cytoplasmic extensions that form an interlacing network throughout the tissue. These cells have long been considered to differentiate into tissue fibroblasts. For many years there has been active debate about the extent to which connective tissue "scar" was the result of an ingrowth of subjacent fibroblast-like elements, or the product of cells derived from the hematogenous entry of circulating, fibroblast precursors.[6] Indeed, the notion that matrix-producing cells could be derived from the peripheral blood has been in the literature for almost 150 years.[6–8] A blood-borne source of "fibroblasts" was first proposed by Cohnheim and discussed further in the writings of Paget, Metchnikov, Fischer, and Maximow.[9–11] While mesenchymal cells have a major role in wound repair, they are not of hematopoietic origin. References in older literature to "blood-borne fibroblasts" and "fibroblast-like cells" exist and indeed may represent the first observations of cells with the current, molecularly-defined, features of circulating fibrocytes. It is likely that in experimental studies of wound repair that go back to at least the 1940s, the cells in the circulating blood that were capable of producing connective tissue represented the first description of "fibrocytes."[7] The studies of Allgöwer for instance, strongly supported the concept of blood-borne cells as the source of new fibroblasts,[12] and investigations by Stirling and Kakkar used cannulation and diffusion chamber methods to demonstrate that collagen-producing cells were not contaminants dislodged from the blood vessel wall, but rather were derived from circulating blood elements.[7]

Circulating fibrocytes were described in 1994 as a unique, $CD34^+$ cell population that produces collagen, and they were identified by virtue of their infiltration from inflammatory exudates into subcutaneously implanted wound chambers.[13] Fibrocytes express collagen and matrix proteins together with cell surface markers indicative of a hematopoietic, bone marrow origin. The name "fibrocyte" was coined by analogy to other circulating cell types, such as erythrocytes, thrombocytes, etc. Fibrocytes thus are unusual, and perhaps unique, in that they are matrix-producing cells that circulate in the peripheral blood. Originally, it was suggested that fibrocytes might have their origin from bone marrow stroma, which is the connective tissue meshwork that functions to support normal hematopoiesis. Both fibrocytes and bone marrow stroma express the CD34 antigen.[14] More recently, studies that have employed green fluorescent protein (GFP)-labeled or sex-mismatched, bone marrow reconstituted mice have provided evidence that fibrocytes are derived from the donor, not host, which may be consistent with a hematopoietic, stem cell origin.[15–18] Recent studies have also demonstrated that fibrocyte outgrowth may be demonstrated in culture from an adherent, $CD14^+$ cell-enriched fraction of peripheral blood, although this may be associated with a time-dependent loss of the CD14 marker.[19,20] Follow-up studies in transwell cultures showed a promotional effect of T cells, or their products in the differentiation or proliferation of fibrocytes from $CD14^+$, peripheral blood precursors.[19] Fibrocyte outgrowth from $CD14^+$ cells has also been reported in studies of cells from human burn patients.[21] Notably, there is precedence for these observations in prior experiments that supported the appearance of collagen-secreting, fibroblast-like cells from cultured, peripheral blood monocyte preparations.[22]

Phenotypic Properties

The original description of circulating fibrocytes relied on the identification of unique, $CD34^+$/collagen+ double-positive cells that were recruited into subcutaneously-implanted wound chambers from an inflammatory exudate.[13] Accordingly, the early phenotypic characterization of fibrocytes from mice relied on the properties of these cells as they were expressed in an "activated" state. In addition, fibrocyte isolation relied on the "self-purification" of these proliferating cells on matrix substratum from a leukocyte-rich (and proliferation-poor), peripheral blood fraction.

Flow cytometric studies identified peripheral blood fibrocytes as expressing the CD34 cell surface antigen. CD34 is a 110kD integral membrane glycoprotein that was initially reported to be expressed exclusively on hematopoietic stem cells, including various myeloid and lymphoid progenitor cells.[14] It is now appreciated that embryonic fibroblasts, endothelial cells, and bone marrow stromal cells also express CD34 to a varying degree.[14,23] Studies in different laboratories have affirmed the utility of the CD34 marker for identifying fibrocytes,[13,24–27] although it is now apparent that the expression of CD34 fibrocyte decreases over time, both in culture and under certain *in vivo* conditions.[15,28–30] Whether this reflects maturation, differentiation (CD34 loss is associated with acquisition of α-smooth muscle expression in some studies),[15,30] or the reversion of an activated phenotype is unclear. It is likely that the *in situ* environment influences the durability of fibrocyte CD34 expression; for example, wound chamber studies in mice have shown an increase in fibrocyte CD34 expression over time.[19] Additional cell surface markers that are indicative of the hematologic origin of fibrocytes may remain stably expressed, such as CD45, HLA-DR, CD71, CD80, and CD86.[13,31] More recent studies have utilized additional markers to identify fibrocytes, such as leukocyte specific protein 1(LSP-1), and Type I procollagen.[15,21] LSP-1 was shown to be stably expressed by fibrocytes after 14 days in culture, and after CD34 had been downregulated.[15] The minimum criterion of the combination of collagen production and uniquely hematologic markers (i.e. CD34, CD45) appears sufficient to describe fibrocytes. Other markers of connective tissue matrix production, such as vimentin, and proline-4-hydroxylase, have been used to identify fibrocytes in hypertrophic scars and keloids.[28]

Of significant biologic interest are the signals that mediate fibrocyte trafficking and recruitment, as these govern the differential role of fibrocytes in various pathologic conditions. Fibrocytes express several receptors for the class of low-molecular weight, chemotactic cytokines, or chemokines, such as CCR3,[19] CCR5,[19] CXCR4,[19,30,32] CCR7,[19,32] and CCR2.[33] Fibrocyte migration into sites of tissue injury can be readily quantified by labeling the cells *ex vivo* with membrane-inserting, fluorescent dyes. The first chemokine/chemokine receptor pair that was described as mediating fibrocyte migration involved SLC (secondary lymphoid chemokine), and CCR7. Simple, intradermal instillation of SLC resulted in the efficient migration of peripheral blood fibrocytes, labeled *ex vivo,* into the injection site.[19] Studies in mice genetically deficient in CCR2 demonstrated the key role of this receptor in the recruitment of fibrocytes to lungs damaged by toxin instillation,[33] and the CXCL12/CXCR4 ligand/receptor pair has been implicated in well-characterized, bleomycin-induced model of acute lung injury.[30]

Functional Roles in Wound Repair

The first functional studies of circulating fibrocytes were in model systems of wound repair. The subcutaneous implantation into mice of wound chambers, which consist of 1.5 cm lengths of polyvinyl-alcohol sponge-filled, silastic tubing induces an exudative response and the rapid recruitment of blood-borne, collagen-producing fibrocytes.[13] In the classic description of wound repair, the host response begins immediately after the traumatic disruption of tissue. Chemoattractants induced by injury to endothelium and other tissue elements lead to the successive recruitment of different cells into the injured site.[1] In the first phase of wound repair, termed the inflammatory phase, neutrophils and monocytes enter into the wound and remove clot, cell debris and invading bacteria. Monocytes mature into macrophages and secrete additional chemoattractants and growth factors: these serve to promote immunity, fibroblast activation and replication, and angiogenesis. In the subsequent proliferative phase, the epithelium covers the wound (in the case of epithelial injury), and fibroblasts actively lay down collagen and other matrix proteins. In the maturation phase, new matrix is slowly matured and remodeled in order to give tensile strength to the injured site and restore the tissue it to its original state of integrity. Specialized cells with contractile ability called myofibroblasts appear and pull the injured wound edges together. Fibrocytes certainly play a role in the proliferative phase of wound repair, and by their early appearance and ability to present antigen, they participate in the first, inflammatory stage of wound healing. Cultured fibrocytes actively produce collagen, albeit not in the high concentrations typified by culture adapted, foreskin fibroblasts.[34] Fibrocytes can be induced by TGF-β to express α-smooth muscle actin, which is a specific marker of the myofibroblast, and they readily contract collagen-gels.[19] In the bronchial tissue of asthmatic lungs, fibrocytes also express α-smooth muscle actin, which contributes to the fibrotic changes evident in chronic asthma.[15]

Fibrocytes also produce angiogenic factors such as vascular endothelial growth factor (VEGF), platelet-derived growth factor A

(PDGF-A), macrophage-colony stimulating factor (M-CSF), hepatocyte growth factor (HGF), granulocyte-macrophage colony stimulating factor (GM-CSF), basic fibroblast growth factor (b-FGF), and connective tissue growth factor (CNTGF).[35] Fibrocytes thus have a potent effect on new blood vessel formation, as evidenced by studies in an *in vivo*, Matrigel model of angiogenesis.[35] Fibrocyte expression of matrix metalloproteinase-9 (MMP-9), which mediates endothelial cell invasion, further facilitates the angiogenic process. Additional pro-angiogenic factors that are present in abundant quantities in fibrocyte conditioned media include interleukin-1β (IL-1β) and interleukin-8 (IL-8).[34,35]

The discovery of significant HLA-DR and co-stimulatory molecule expression by fibrocytes prompted a systematic study of their antigen presenting properties *in vitro* and *in vivo*.[31] Fibrocytes stimulate human tetanus toxoid-specific T cell responses by measures that are superior to peripheral blood monocytes and equivalent to that of dendritic cells, and they have the ability to take-up and present bacterially-encoded antigens[31,36] (unpublished observations). Mouse studies confirmed the ability of antigen-primed fibrocytes to migrate to regional lymph nodes and to prime naïve T cells in an antigen- and class II-dependent fashion.[31] Fibrocytes also prime CD8 T cell responses.[36] The precise role of fibrocytes in lesional antigen presentation, *vis-a-vis* the other antigen-presenting cells present in inflammatory lesions remains to be elucidated. Fibrocytes nevertheless remain present in granulomas,[34] and likely contribute both to the chronic immunostimulatory milieu of the lesions and to the fibrogenic encapsulation response.

Role in Health and Disease

The clinical relevance of fibrocytes centered first on their role in wound repair and in inflammatory fibrosis, as typified by the granuloma.[13,34] As developed in this volume, there is increasing evidence for a role for fibrocytes in hypertrophic scars, tumors, scleroderma and related disorders, and pulmonary fibrosis.

Wound Repair

In cutaneous wounds, especially in the expanding margins of keloids or inflammatory scars, fibrocyte CD34 expression decreases over time while the expression of proline-4-hydroxylase, an enzyme necessary for the production of mature collagen increases.[28,37] The reversion of CD34 expression in scars may represent a decrease in the "inflammatory" state of the wound, and that overall CD34 positivity is inversely proportional to collagen synthesis. Locally produced IL-1, which is a major inflammatory cytokine in tissue injury, likely regulates the phenotypic transition in fibrocyte function from an inflammatory to a remodeling phase.[34] High levels of IL-1, which are present in the early, inflammatory stages of tissue injury, promote fibrocyte proliferation, spreading, and the acquisition of a spindle-shaped morphology. IL-1 also promotes fibrocyte expression of the cytokines TNF-α, IL-6, IL-8, IL-10, and chemokines (e.g. MIP-1α/β), and the secretion of matrix metalloproteases (MMP-9), while suppressing collagen production.[34] As inflammatory levels of IL-1 decrease, this phenotypic transition is likely augmented by the increasing levels of TGF-β present in the lesions, which plays an important role in inducing the synthesis of collagen and α-smooth muscle actin.[9] TGF-β stimulation leads to a concomitant decrease in cell surface CD34 expression,[16,19] and investigations in lung tissue have confirmed that the loss of CD34 expression is associated with an increase in α-smooth muscle actin production.[15] The acquisition of α-smooth muscle actin expression is considered to be a differentiating characteristic of the contractile myofibroblast. This transition of fibrocytes to a myofibroblast phenotype thus contributes to the progression of many types of pathologic fibroses.[38]

The reported difficulty in maintaining stable CD34 expression in fibrocyte lines derived from different tissue sources thus may be due to the downregulation of this gene after the exit of fibrocytes from the circulation, and to their subsequent maturation into a more differentiated connective tissue cell type. In the hypertrophic scars that arise as a result of burn injury, fibrocyte numbers have been monitored by measurements of collagen I together with the specific, fibroblast marker LSP-1.[21]

Tumor Biology

Several histopathologic studies have correlated the loss of $CD34^+$ stromal cells in tumor sites with malignant potential. It is likely that these cells are fibrocytes.[24] We and others also have noted the encapsulation of benign tumors by $CD34^+$ connective tissue cells[24,39] (unpublished observations). The tumors include breast, skin, pancreatic, and cervical cancer.[24–26,39–41] In the case of breast cancer, benign lesions show significant numbers of $CD34^+$ cells in the stroma, whereas examination of ductal carcinoma *in situ* (a pre-malignant state) and invasive breast cancer revealed a marked loss of $CD34^+$ cells.[25] In pancreatic cancer, ductal adenocarcinoma and endocrine tumors show a paucity of $CD34^+$ cells, whereas in chronic pancreatitis, the $CD34^+$ cell numbers are maintained.[26] Invasive carcinoma of the cervix also can be differentiated from cervical intraepithelial neoplasia by the loss of CD34 positivity.[24] Basal cell carcinoma of the skin and colorectal adenoma are two additional examples of malignancies where the loss of $CD34^+$ cells may have prognostic relevance. It may be hypothesized that metastatic progression is accompanied by the loss of a host fibrocyte response to the invasive tumor, which may be necessary for a successful encapsulation response.

Immunostimulatory Properties

The potent antigen-presentation properties of fibrocytes together with their facile propagation from the peripheral blood prompted a phase I clinical trial of autologous fibrocytes for the treatment of metastatic cancer.[31,44] The rationale for this approach followed that of cell-based, immunotherapies utilizing professional "antigen-presenting cells" such as dendritic cells. To the extent that cancer is associated with an inappropriate immunosuppressive response by the host, antitumor immunity may be augmented by the delivery of antigen presenting cells loaded with tumor antigens. Studies in tumor bearing mice (fibrosarcoma or melanoma) validated this approach for a fibrocyte, cell-based immunotherapy. In the human protocol, fibrocytes were isolated from patients by leukophoresis and their numbers expanded by *in vitro* cultivation. Cultures then were pulsed with an autologous tumor lysate obtained by biopsy, and the antigen-loaded fibrocytes re-administered to each patient by subcutaneous injection in a dose-escalating schedule. While the initial phase I trial was not designed to demonstrate clinical efficacy, it did support the fundamental safety of the administration of fibrocytes after *in vitro* expansion and conditioning.[44] Studies in primate models of AIDS have also explored the transfectability of fibrocytes as a means to deliver foreign antigens to counter host immunosuppression or defective antigen presentation.[45]

Infectious Diseases

The causative pathogen of Lyme disease, the spirochete *Borrelia burgdorferi*, binds to both human and monkey (rhesus) fibrocytes by a process that does not require the well-characterized OspA or OspB borrelial surface proteins.[46] The spirochetes are not phagocytosed but instead are taken into deep recesses of the cell membrane by coiling phagocytosis, which may shield them from immune destruction. Grab and colleagues have proposed that this interaction between *B. burgdorferi* and peripheral blood fibrocytes may explain the targeting of spirochetes to connective tissue and contribute to the inflammatory process in Lyme arthritis.[47] Circulating fibrocytes may under go infection by "foamy" viruses (*Retroviridae* sub-group *Spumavirinae*), although the clinical significance of this finding remains to be determined.[48] A role for fibrocytes in directing the adaptive immune response in the parasitic infection *Leishmaniasis* has also been proposed, based on the characteristic expression of T_H2 cytokines, TGF-β, and CD86 by these cells.[49]

Scleroderma

Scleroderma is an autoimmune disease of unknown etiology that is distinguished clinically by restrictive fibrosis and an obliterative vasculopathy.[50] The inflammatory response in skin and other organs is characterized by the persistence of immune cells (macrophages, lymphocytes) and α-smooth actin positive cells, leading to irreversible fibrosis of the skin and major organs. A progressive vasculopathy results in significant tissue and end-organ complications, and it is characterized pathologically by smooth muscle cell proliferation within the vessel wall and the obliteration of the normal luminal diameter. The fibrosis in scleroderma results from an increase in connective tissue cells and the production of matrix proteins including collagen. An examination of specimens from 27 scleroderma patients revealed few $CD34^+$ cells in the skin as compared to healthy controls or patients with other collagen vascular diseases[29] (Quan *et al.*, personal communication). The paucity of CD34 immunoreactivity in the fibrotic areas of scleroderma lesions may reflect the same progression of events that has been described for wound repair, where CD34 expression is prominent early in the inflammatory areas but then downregulated over time.[29] A recent study has described a regulatory role for serum amyloid P in fibrocyte differentiation, and there may exist a subset of patients in whom circulating serum amyloid P levels influence circulating fibrocyte numbers and pathogenic potential.[20]

Nephrogenic Systemic Fibrosis (NSF)

NSF is a newly emerging fibrosing disease of the skin that is clinically similar to myxedema and scleroderma.[51] The disease is strongly associated with renal insufficiency, and while its cause remains an enigma, an environmental or treatment-related etiology is suspected. Concurrent hypercoagulability, surgery, and liver disease have been proposed as precipitating factors.[52,53] The skin becomes thickened and "woody" and while the disease is usually limited to the extremities, the speed with which the skin lesions develop can be remarkable. The pathognomonic feature is the presence of dual positive CD34/procollagen spindled-shaped cells (i.e. fibrocytes) in a thickened dermis.[54] The relative absence of mitotic spindles has prompted the hypothesis that NSF is a disease of enhanced fibrocyte activation or trafficking to skin. Whether NSF is truly a primary disorder of fibrcoytes, or of fibrocyte trafficking, remains to be determined. The histologic prominence of fibrocytes in NSF affected skin, when contrasted with scleroderma skin, may reflect the rapid onset of disease in NSF versus the more indolent course of disease in scleroderma patients.

Asthma, Acute Lung Injury, and Pulmonary Fibrosis

Recent studies have opened new avenues of inquiry based on evidence that circulating fibrocytes contribute to the development of pulmonary fibrosis. The seminal study in this series provided both experimental animal and human data for a role for fibrocytes in bronchial asthma.[15] In patients with allergen-induced asthma, endobronchial biopsies performed after antigen challenge showed that airway fibrosis was associated with the presence of $CD34^+$/collagen+ fibrocytes. Notably, there also was an inverse correlation, over time, between $CD34^+$ expression and the appearance of α-smooth muscle actin expressing myofibroblasts. In a complementary mouse model of the disease, fibrocytes were labeled, reintroduced into the circulation, and observed to traffic into the bronchial tissue of the lung at the time of asthma induction. It was found that under the influence of TGF-β or endothelin-1, fibrocytes downregulated CD34, upregulated α-smooth muscle actin, and differentiated into myofibroblasts.[15] In the bleomycin model of acute lung injury, it was demonstrated that fibrocytes from donor bone marrow entered the lung and produced connective tissue matrix.[30] In a subsequent study, the chemokine receptor CCR2 was found to be crucial to the trafficking of fibrocytes into lungs damaged by toxin instillation.[33]

Conclusions

Our current molecular definition of fibrocytes by gene expression and flow cytometry studies has brought to fruition many years of observations, some made as far back as the mid-19th century, which have supported a blood-borne source for a connective tissue cell population.[7–9,11,12] Studies performed over the last 10 years have provided solid support for the fibrocyte as a collagen-producing progenitor cell of the peripheral blood. The accumulated evidence indicates that circulating fibrocytes play an important role in the inflammatory and proliferative phases of the host response to injury or tissue invasion. This is supported by numerous observations of fibrocyte migration into sites of injury, and by their regulated production of inflammatory and growth regulating cytokines, and matrix proteins. Fibrocytes are cellular constituents of granulomas, healing wounds and hypertrophic scars, and the bronchial lesions of asthma.

Based on accumulated information, we propose a model where damage from wounding, inflammation, or invasive tumors recruits fibrocytes from the peripheral circulation (Fig. 1). These fibrocytes originate at least in part from bone marrow derived cells including circulating $CD14^+$ precursor cells. Circulating fibrocytes express CD34, but they downregulate the expression of this surface protein as they become more specialized. The reduction in the cell surface expression of CD34 antigen by circulating fibrocytes both *in vitro* and in several *in vivo* contexts likely reflects α-smooth muscle actin expression, terminal differentiation, or other phenomena specific to a particular tissue microenvironment. Mediators such as TGF-β,

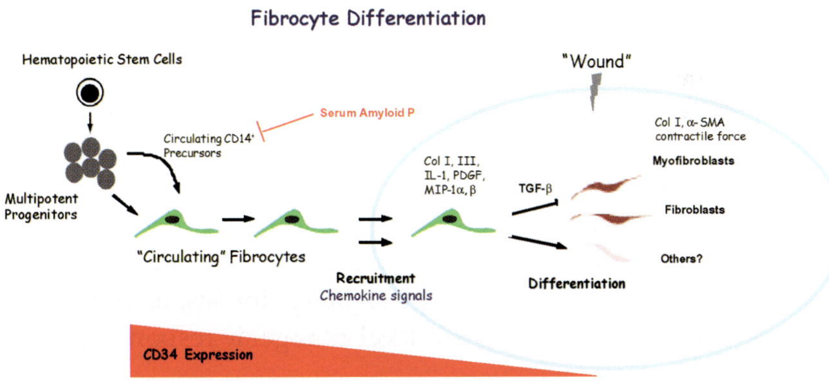

Fig. 1. Model for fibrocyte function and differentiation.
Tissue damage from various causes leads to the recruitment of fibrocytes from the peripheral circulation. Circulating fibrocytes originate in part from $CD14^+$ precursor cells from the bone marrow. Circulating fibrocytes are $CD34^+$ but lose the expression of this antigen as they take up residence in tissue and become more differentiated (*modified from* Quan *et al.* 2006).

IL-1, SLC, CXCL12, and serum amyloid P influence fibrocyte function, proliferative potential, differentiation into myofibroblasts, and trafficking properties. Whether the circulating fibrocyte explains the origin of the myofibroblast, which features prominently in many pathologic fibroses, is of interest in better understanding many chronic diseases. This model of fibrocyte action does not discount the important role of locally-derived, connective tissue cells in the host response but suggests that the contribution of circulating fibrocytes to a particular site of injury likely reflects the inflammatory character of the lesion and the relative abundance of fibrocytes, which may enter from exudative fluid, versus local connective tissue sources of fibroblasts.

Future studies will need to better define the fundamental properties of fibrocytes, and their differentiation potential in different inflammatory and tissue injuries. The recent description of NSF has served to focus significant attention on the pathologic role of fibrocytes. Whether NSF is a disease of aberrant fibrocyte activation, trafficking, or synthetic function remains unknown. A better understanding of the mediators and the signaling pathways that govern fibrocyte trafficking and differentiation under different circumstances may lead not only to a better understanding of fibrosing disorders such as NSF, but may suggest means to augment fibrocyte function for therapeutic benefit. The ability to expand fibrocytes *ex vivo* for therapeutic re-administration may also prove to be of clinical utility, as has been suggested by one clinical study of fibrocytes for immunotherapy in metastatic cancer. Ultimately, deeper knowledge of the molecular and cellular biology of fibrocytes may aid in the unraveling of common fibrosing disorders such as idiopathic pulmonary fibrosis, or conditions such as atherosclerosis, in which local fibrogenic response plays a critical role.

Acknowledgments

These studies were supported by the NIH, the Scleroderma Foundation, the American College of Rheumatology, and the Yale General Clinical Research Center.

References

1. Clark RAF. (1996) Wound repair: overview and general considerations. In: Clark RAF (ed.), *The Molecular and Cellular Biology of Wound Repair*, Plenum Press, New York, pp. 3–50.
2. Homer RJ, Elias JA. (2000) Consequences of long-term inflammation — airway remodeling. *Clin Chest Med* **21**: 331–343.
3. Liu YH. (2006) Renal fibrosis: new insights into the pathogenesis and therapeutics. *Kidney Int* **69**: 213–217.
4. Coussens LM, Werb Z. (2002) Inflammation and cancer. *Nature* **420**: 860–867.
5. Orimo A, Gupta PB, Sgroi DC, *et al.* (2005) Stromal fibroblasts present in invasive human breast carcinomas promote tumor growth and angiogenesis through elevated SDF-1/CXCL12 secretion. *Cell* **121**: 335–348.
6. Dunphy J. (1963) The fibroblast — a ubiquitous ally for the surgeon. *New Engl J Med* **268**: 1367–1377.
7. Stirling GA, Kakkar VV. (1969) Cells in the circulating blood capable of producing connective tissue. *Br J Exper Pathol* **50**: 51–55.
8. Paget J. (1863), *Lectures on Surgical Pathology*, Longmans, London, pp. 1–848.
9. Cohnheim J. (1867) Ueber Entzundung und Eiterung (About inflammation and suppuration). *Path Anat Physiol Klin Med* **40**: 1–79.
10. Fischer A. (1925) *C.r. Seanc Soc Biol* **92**: 109.
11. Maximow AA. (1928) *Arch exp Zellforsch* **5**: 169.
12. Allgower M, Hulliger L. (1960) Origin of fibroblasts from mononuclear blood cells: A study on *in vitro* formation of the collagen precursor, hydroxyproline, in buffy coat cultures. *Surgery* **47**: 603–609.
13. Bucala R, Spiegel L, Chesney J, *et al.* (1994) Circulating fibrocytes define a new leukocyte subpopulation that mediates tissue repair. *Mol Med* **1**: 71–81.
14. Brown J, Greaves M, Molgaard H. (1991) The gene encoding the stem cell antigen, CD34, is conserved in mouse and expressed in hematopoietic progenitor cell lines, brain, and embryonic fibroblasts. *Int Immunol* **3**: 175–184.
15. Schmidt M, Sun G, Stacey MA, *et al.* (2003) Identification of circulating fibrocytes as precursors of bronchial myofibroblasts in asthma. *J Immunol* **171**: 380–389.
16. Mori L, Bellini A, Stacey MA, *et al.* (2005) Fibrocytes contribute to the myofibroblast population in wounded skin and originate from the bone marrow. *Exper Cell Res* **304**: 81–90.

17. Fathke C, Wilson L, Hutter J, *et al.* (2004) Contribution of bone marrow-derived cells to skin: collagen deposition and wound repair. *Stem Cells* **22**: 812–822.
18. Opalenik SR, Davidson JM. (2005) Fibroblast differentiation of bone marrow-derived cells during wound repair. *FASEB J* **19**: 1561–1563.
19. Abe R, Donnelly SC, Peng T, *et al.* (2001) Peripheral blood fibrocytes: differentiation pathway and migration to wound sites. *J Immunol* **166**: 7556–7562.
20. Pilling D, Buckley CD, Salmon M, Gomer RH. (2003) Inhibition of fibrocyte differentiation by serum amyloid P. *J Immunol* **171**: 5537–5546.
21. Yang LJ, Scott PG, Giuffre J, *et al.* (2002) Peripheral blood fibrocytes from burn patients: identification and quantification of fibrocytes in adherent cells cultured from peripheral blood mononuclear cells. *Lab Invest* **82**: 1183–1192.
22. Labat ML, Bringuier AF, Arysphilippart C, *et al.* (1994) Monocytic origin of fibrosis — *in vitro* transformation of Hla-Dr monocytes into neo-fibroblasts — inhibitory effect of all-trans-retinoic acid on this process. *Biomed Pharmacother* **48**: 103–111.
23. Fina L, Molgaard H, Robertson D, *et al.* (1990) Expression of the CD34 gene in vascular endothelial cells. *Blood* **75**: 2417–2426.
24. Barth PJ, Ramaswamy A, Moll R. (2002a) CD34($^+$) fibrocytes in normal cervical stroma, cervical intraepithelial neoplasia III, and invasive squamous cell carcinoma of the cervix uteri. *Virchows Archiv* **441**: 564–568.
25. Barth PJ, Ebrahimsade S, Ramaswamy A, Moll R. (2002b) CD34$^+$ fibrocytes in invasive ductal carcinoma, ductal carcinoma *in situ*, and benign breast lesions. *Virchows Archiv* **440**: 298–303.
26. Barth PJ, Ebrahimsade S, Hellinger A, *et al.* (2002c) CD34$^+$ fibrocytes in neoplastic and inflammatory pancreatic lesions. *Virchows Archiv* **440**: 128–133.
27. Hirohata S, Yanagida T, Nagai T, *et al.* (2001) Induction of fibroblast-like cells from CD34($^+$) progenitor cells of the bone marrow in rheumatoid arthritis. *J Leukocyte Biol* **70**: 413–421.
28. Aiba S, Tagami H. (1997) Inverse correlation between CD34 expression and proline-4-hydroxylase immunoreactivity on spindle cells noted in hypertrophic scars and keloids. *J Cutaneous Pathol* **24**: 65–69.
29. Aiba S, Tabata N, Ohtani H, Tagami H. (1994) CD34$^+$ spindle-shaped cells selectively disappear from the skin lesion of scleroderma. *Arch Dermatol* **130**: 593–597.

30. Phillips RJ, Burdick MD, Hong K, *et al.* (2004) Circulating fibrocytes traffic to the lungs in response to CXCL12 and mediate fibrosis. *J Clin Invest* **114**: 438–446.
31. Chesney J, Bacher M, Bender A, Bucala R. (1997) The peripheral blood fibrocyte is a potent antigen-presenting cell capable of priming naive T-cells *in situ*. *Proc Natl Acad Sci USA* **94**: 6307–6312.
32. Hashimoto N, Jin H, Liu TJ, *et al.* (2004) Bone marrow-derived progenitor cells in pulmonary fibrosis. *J Clin Invest* **113**: 243–252.
33. Moore BB, Kolodsick JE, Thannickal VJ, *et al.* (2005) CCR2-mediated recruitment of fibrocytes to the alveolar space after fibrotic injury. *Am J Pathol* **166**: 675–684.
34. Chesney J, Metz C, Stavitsky A, *et al.* (1998) Regulated production of type I collagen and inflammatory cytokines by peripheral blood fibrocytes. *J Immunol* **160**: 419–425.
35. Hartlapp I, Abe R, Saeed RW, *et al.* (2001) Fibrocytes induce an angiogenic phenotype in cultured endothelial cells and promote angiogenesis. *FASEB J* **15**: 2215–2224.
36. Balmelli C, Ruggli N, McCullough K, Summerfield A. (2005) Fibrocytes are potent stimulators of anti-virus cytotoxic T cells. *J Leukocyte Biol* **77**: 923–933.
37. Yang LJ, Scott PG, Dodd C, *et al.* (2005) Identification of fibrocytes in postburn hypertrophic scar. *Wound Repair Reg* **13**: 398–404.
38. Gabbiani G. (2003) The myofibroblast in wound healing and fibrocontractive diseases. *J Pathol* **200**: 500–503.
39. Chauhan H, Abraham A, Phillips JRA, *et al.* (2003) There is more than one kind of myofibroblast: analysis of CD34 expression in benign, *in situ*, and invasive breast lesions. *J Clin Pathol* **56**: 271–276.
40. Ramaswamy A, Moll R, Barth PJ. (2003) $CD34^+$ fibrocytes in tubular carcinomas and radial scars of the breast. *Virchows Archiv* **443**: 536–540.
41. Kirchmann TTT, Prieto VG, Smoller BR. (1994) CD34 staining pattern distinguishes basal-cell carcinoma from trichoepithelioma. *Arch Dermatol* **130**: 589–592.
42. Barth PJ, Schweinsberg TSZ, Ramaswamy A, Moll R. (2004) $CD34^+$ fibrocytes, alpha-smooth muscle antigen-positive myofibroblasts, and CD117 expression in the stroma of invasive squamous cell carcinomas of the oral cavity, pharynx, and larynx. *Virchows Archiv* **444**: 231–234.
43. Naftzger C, Mule JJ, Bucala R, Rice G. (1998) Cell process optimization of fibrocytes, a novel antigen presenting cell. Society for Biological Therapy.

44. Naftzger C, Mule J, Redman B, *et al.* (1998) Fibrocytes are a novel antigen presenting cell for autologous anti-cancer therapy. Cancer Vaccine Meeting, NY.
45. Zhu YD, Koo K, Bradshaw JD, *et al.* (2000) Macaque blood-derived antigen-presenting cells elicit SIV-specific immune responses. *J Med Primatol* **29**: 182–192.
46. Grab DJ, Lanners HN, Martin LN, *et al.* (1999) Interaction of *Borrelia burgdorferi* with peripheral blood fibrocytes, antigen-presenting cells with the potential for connective tissue targeting. *Molec Med* **5**: 46–54.
47. Grab DJ, Salim M, Chesney J, *et al.* (2002) A role for peripheral blood fibrocytes in Lyme disease? *Med Hypoth* **59**: 1–10.
48. Grab DJ, Lanners HN, Williams WL, Bucala R. (1999) Peripheral blood fibrocytes with foamy virus infection-like morphology. *Hum Pathol* **30**: 1395–1396.
49. Grab DJ, Salem ML, Dumler JS, Bucala R. (2004) A role for the peripheral blood fibrocyte in leishmaniasis? *Trends Parasitol* **20**: 12.
50. White B. (2001) Systemic sclerosis and related syndromes, epidemiology, pathology, and pathogenesis. In: Klippel J (ed.), *Primer on the Rheumatic Diseases*, Arthritis Foundation, Atlanta, pp. 354–357.
51. Cowper S, Su L, Bhawan J, *et al.* (2001) Nephrogenic fibrosing dermopathy. *Am J Dermatopathol* **23**: 383–393.
52. Cowper SE, Bucala R, Leboit PE. (2005) Case 35-2004: Nephrogenic fibrosing dermopathy. *N Engl J Med* **352**: 1723.
53. Cowper SE. (2003) Nephrogenic fibrosing dermopathy: the first 6 years. *Curr Opin Rheumatol* **15**: 785–790.
54. Cowper SE, Bucala R. (2003) Nephrogenic fibrosing dermopathy: suspect identified, motive unclear. *Am J Dermatopathol* **25**: 358.
55. Quan T, Cowper S, Bucala R. (2006) The role of circulating fibrocytes in fibrosis. *Curr Rheumatology Rep* **8**: 145–150.

Chapter 2

Fibrocytes: Immunologic Features

Jason Chesney*

Introduction

Fibrosis is a frequent pathological sequela of several common diseases that cause significant morbidity and mortality, including atherosclerosis, interstitial lung disease, cirrhosis, macular degeneration, glomerulosclerosis and scleroderma. Fibrosis is caused by fibroblasts that secrete extracellular matrix proteins such as types I and III collagens. Accumulation of these matrix proteins replaces normal organ parenchyma, leading to their dysfunction and ultimate failure. Unfortunately, once the fibrotic response is initiated, therapeutic options to disrupt the progression are limited. For example, the only efficacious treatment for life-threatening cirrhosis is liver transplantation. Accordingly, there is a great need for an increased understanding of the cellular and humoral events that both initiate and propagate collagen deposition. This chapter will summarize the significance of T lymphocyte infiltration into fibrogenic tissues

*Division of Hematology/Oncology, Department of Medicine, J.G. Brown Cancer Center, University of Louisville, 529 South Jackson Street, Rm# 425, Louisville, Kentucky 40202, USA. Tel: (502)852–3402; Fax: (502) 852-5679; Email: jason.chesney@louisville.edu

and the role of collagen-secreting peripheral blood fibrocytes in their migration and antigen-specific activation.

Requirement of T Cells for the Development of Fibrosis

Although the precise etiologies of most fibrotic disorders are poorly understood, the immune cells that infiltrate normal tissue prior to the deposition of collagens have been well defined. Initially, neutrophils and macrophages dominate, followed by an infiltration of lymphocytes thought to be secondary to the secretion of chemoattractants. The presence of T lymphocytes within the areas of injury that ultimately become scarred support an essential role for T lymphocytes in the progression of inflammation to fibrosis.[1] Direct evidence for the role of $CD4^+$ and $CD8^+$ T lymphocytes is derived from mouse models of fibrosis in which anti-CD4 or anti-CD8 depleting monoclonal antibodies have been found to markedly attenuate fibrotic progression. Mice exposed to trinitrophenyl develop pulmonary interstitial fibrosis within seven days of administration and a marked T lymphocyte infiltrate is observed within the fibrotic lung tissue.[2] Inflammatory and fibrotic sequelae can be adoptively transferred with lymphocytes from sensitized mice and administration of anti-CD4 or anti-CD8 monoclonal antibodies to sensitized mice attenuates collagen deposition.[2] A similar reduction in fibrosis has been observed with anti-CD4 administration to mice suffering from pulmonary fibrosis caused by intratracheal silica instillation.[3]

The requirement of T lymphocytes for the development of fibrotic disorders has also been studied using athymic mice that are unable to produce T lymphocytes. Pathological fibrosis is reduced in athymic mice that have been inflicted with several fibrotic disorders, including interstitial renal fibrosis[4]; pulmonary fibrosis caused by silica particles[5] or bleomycin[6,7]; and hepatic fibrosis mediated by group A streptococcal cell walls.[8] Peri-tumor fibrosis is an essential requirement for tumor growth, and collagen deposition associated mammary carcinoma growth is reduced in athymic mice.[9] In summary, the observations that either monoclonal antibody depletion of T lymphocytes or the congenital absence of a thymus are sufficient to attenuate fibrosis initiated by different stimuli in disparate organs, support an essential role for T lymphocytes in the fibroses of human diseases.

T Cell-mediated Fibrosis in Human Disease

In the United States, about 16 million people suffer from chronic obstructive pulmonary disease (COPD).[10] It is the fourth most common cause of death, accounting for more than 100,000 deaths per year in the United States; the number of deaths from COPD has increased by 40% over the last 20 years.[10] In COPD an inflammatory reaction followed by normal tissue destruction and fibrosis is thought to be the major cause of the development of airway abnormalities and emphysema, and consequently COPD, in susceptible smokers. The earliest pathological abnormality in the airway of smokers is a cellular inflammatory infiltrate throughout the wall. Like most fibrotic disorders, the stimuli for this inflammatory infiltrate are not known, but it is possible that injury to the airway epithelium, which is the first structure encountered by cigarette smoke, promotes and perpetuates inflammation in the airways. Saetta *et al.* recently investigated the differences in airway inflammation in smokers, who either developed COPD or did not, by examining surgical specimens obtained from the following two groups of smokers: asymptomatic smokers with normal lung function, and symptomatic smokers with COPD.[11] While both groups were of similar age and had similar smoking histories, smokers with COPD differed in the nature of their inflammatory responses in the small airways. Smokers with COPD had increased numbers of $CD8^+$ T lymphocytes in the walls of the small airways compared with the healthy smokers. Other cells, including neutrophils, were similar in number in the two groups of smokers. Interestingly, not only were $CD8^+$ T cells more prevalent in COPD patients, but the number of these cells increased with worsening airflow limitation.[11] Similar findings have been reported by O'Shaughnessy *et al.*, who demonstrated an increased number of $CD8^+$ T cells in bronchial biopsy specimens obtained by bronchoscopy in subjects with COPD when compared with those from smokers without COPD.[12]

The direct correlation between T lymphocytes and fibrosis in human disease is certainly not limited to COPD. $CD4^+$ T lymphocytes have been found in human atherosclerotic plaques and approximately 10% of the T cells cloned from these lesions recognize pro-atherosclerotic oxidized low-density lipoproteins.[13–15] Similarly, T lymphocytes have been observed in experimental models of atherosclerosis, including the apoE knockout mice.[16] $ApoE^{-/-}$ immunodeficient *scid/scid* mice develop atherosclerosis but at a markedly reduced rate relative to $ApoE^{-/-}$ immunocompetent mice, and adoptive transfer of $CD4^+$ T lymphocytes from $apoE^{-/-}$ mice into $apoE^{-/-}$ *scid/scid* mice enables the normal development of atherosclerosis.[16,17] T lymphocytes have also been observed to have infiltrated fibrotic tissues resected from patients with idiopathic pulmonary fibrosis,[18] renal allograft fibrosis[19] and viral hepatitis-induced fibrosis.[20]

$CD4^+$ or $CD8^+$ T cells have been directly correlated with the extent of fibrosis in several human diseases. In experimental models of fibrosis, depletion of these lymphocytes attenuates fibrotic responses, and adoptive transfer of lymphocytes isolated from mice with fibrotic organs accelerates fibrosis. Taken together, these data support an essential role for T lymphocytes in the initiation and progression of fibrosis.

Fibrocytes

Wounding is defined as a physical disruption of the normal architecture of tissue and may be caused by trauma, burns, inflammatory processes or metabolic insufficiency. The host initiates a coordinated repair response to wounding that serves to prevent infection and ultimately reestablish normal tissue integrity. Neutrophils and monocytes function to clear foreign particles, including microorganisms, by phagocytosis and by the release of radical species and bactericidal proteins. These cells also secrete cytokines that serve to combat infection and coordinate successive steps of the tissue repair response.[21,22] Connective tissue cells such as fibroblasts and endothelial cells infiltrate the injured site, proliferate and secrete collagens and other extracellullar matrix proteins essential for the repair phase of wound healing. The usual outcome of this cascading series of cellular events is elimination of the invasive stimulus, followed by connective tissue scar formation and, over time, by remodeling of the injured site. Unfortunately, in diseases of chronic inflammation, fibrosis can consume entire organs.

Investigations into the cell population present in experimentally implanted fibrogenic wound chambers led to the discovery of an adherent, proliferating cell type that displayed fibroblast properties yet expressed distinct hematopoietic/leukocyte cell surface markers.[23] Wound chambers consist of short lengths of sponge-filled, silastic tubing and are a frequently employed model for the study of tissue reparative and fibrotic responses *in vivo*. Implantation of these chambers into the subcutaneous space of mice results in a rapid infiltration of peripheral blood inflammatory cells, including neutrophils, monocytes and lymphocytes.[24] Large numbers of adherent, spindle-shaped cells that resemble fibroblasts were unexpectedly observed to infiltrate wound chambers soon after implantation and coincidentally with the appearance of circulating inflammatory cells.[23] Double immunofluorescence studies showed that within 24 hours of implantation, as many as 10–15% of the cells present in the wound chamber fluid stain positively both for type I collagen and for CD34.[23] These studies suggested the presence in wounds of a previously uncharacterized cell type displaying fibroblast-like features but expressing markers for bone marrow-derived cells. Follow-up immunohistochemical analysis of wound chambers that had been implanted in mice confirmed the presence of $CD34^+$ spindle-shaped cells in areas of collagen matrix deposition.

Termed fibrocytes, these cells can be isolated from blood by employing culture conditions that are selective for fibroblast growth (DMEM, 20% FCS) and that rely on two general properties of the primary fibroblast, its ability to adhere to plastic and to proliferate *in vitro*.[23] Peripheral blood fibrocytes constitutively express in culture the fibroblast products type I collagen, type III collagen and fibronectin, as well as the leukocyte common antigen CD45RO, the pan-myeloid antigen CD13, and the hematopoietic stem cell antigen CD34.[23] Fibrocytes do not synthesize epithelial (cytokeratin), endothelial (von Willebrand factor VIII-related protein), or smooth muscle (α-actin) cell markers and are negative for non-specific esterases as well as the monocyte/macrophage-specific markers, CD14 and CD16.[23] Fibrocytes also do not express the Langerhans cell marker CD1a (data not shown), proteins produced by dendritic cells or their precursors (CD25, CD10, and CD38), or the pan-B cell antigen CD19.[23] Scanning electron microscopy has shown these cells to be morphologically distinct from blood-borne leukocytes and to display unique cytoplasmic extensions intermediate in size between microvilli and pseudopodia.[23]

Fibrocytes are Potent Antigen Presenting Cells

In order for antigen-specific immunity to be initiated, naive $CD4^+$ T cells must be induced by antigen presenting cells (APCs) to proliferate and differentiate into "primed" memory-type T cells that can carry out various effector functions. The sensitization of naive T cells had been considered to be a unique function of dendritic cells,[25,26] although blood-borne B cells had been reported to exhibit this activity *in vitro*.[27] Steinman *et al.* have shown that dendritic cells isolated from mouse spleen and injected subcutaneously homed to draining lymph nodes (1–2% of total injected) and primed naive T cells *in situ*.[25,28] However, other "professional" APCs such as peritoneum-derived macrophages and spleen-derived B cells were found to be incapable of priming naive T cells *in situ*.[25]

Antigen presenting cells express cell surface proteins that form a synapse with T cells and enhance stimulation in response to the MHC-peptide complex. Specific receptor: co-receptor pairs that have been found to be essential include CD11a/18-CD54, CD58-CD2 and CD86-CD28. Peripheral blood fibrocytes express the adhesion molecules, CD11a, CD54 and CD58 at similar levels to that observed in peripheral blood monocytes.[29] Interestingly, fibrocyte expression of the class II MHC molecules, HLA-DP and HLA-DQ, is higher in fibrocytes than even monocytes, although the expression of HLA-DR is equivalent in both cell types.[29] Fibrocytes also express the co-stimulatory molecule CD86 (B7-2) at a level similar to that of monocytes.[29]

Since fibrocytes express the cell surface machinery required to activate T lymphocytes, their ability to induce antigen-dependent T lymphocyte proliferation has been examined. Perhaps not surprisingly, fibrocytes were found to be potent activators of antigen-specific $CD4^+$ T lymphocytes.[29] Importantly, the antigen-dependent T lymphocyte proliferation response induced by monocytes was significantly lower than that induced by fibrocytes but the antigen-dependent T lymphocyte proliferation induced by dendritic cells was higher than that induced by fibrocytes.[29] However, more dendritic cells than fibrocytes were required to achieve this level of proliferative activity.[29] Fibrocyte-induced T lymphocyte proliferation exhibited classical surface protein requirements as proliferation was inhibited significantly by neutralizing monoclonal antibodies to HLA-DR, CD86, CD11a or CD54.[29] The importance of fibrocyte antigen presentation *in vivo* was supported by the observation that subdermal regions of cutaneous scar tissues contain numerous connective tissue matrix-associated inflammatory cells that express both CD34 and HLA-DR, a distinct phenotype of fibrocytes.[29]

Although several cell types have been shown to be capable of presenting antigen to memory T cells, the priming of naive T cells had been considered to be a specialized function of "professional" APCs, particularly dendritic cells.[25,26] Mouse fibrocytes pulsed with the HIV proteins p24 or gp120 *in vitro* and injected intradermally into the rear foot pad of unprimed naïve BALB/c mice were sufficient to induce a strong T cell proliferative response which was specific for the priming antigen (p24 or gp120) and consisted predominantly of $CD4^+$ T lymphocytes.[29]

That antigen-pulsed fibrocytes were not simply transferring antigen to other host APC types was established by experiments in which antigen-pulsed fibrocytes from two parent mouse strains were injected into F_1 offspring mice. The T cell reactivity of F_1 offspring is confined predominantly to antigens presented by one of the parental strains, and the priming and re-stimulation APCs must necessarily share the same haplotype.[25,30] When a parental strain was used as the source of re-stimulation APCs, the F_1 APC-depleted lymph node cells would only proliferate if the priming fibrocytes were from the same parental strain.[29] These data indicate that fibrocyte priming and APC re-stimulation of sensitized T cells occur only in the setting of a shared MHC haplotype, and fibrocytes thus function to directly sensitize naive T cells in a MHC-specific manner. Finally, fluorescently-labeled mouse fibrocytes were found to migrate from the rear footpads of mice into draining popliteal lymph nodes where antigen presentation occurs predominantly.[29]

Fibrocytes isolated from peripheral blood have been shown to: (1) express several cell surface proteins required for antigen presentation; (2) be potent stimulators of antigen-specific T cells *in vitro*; (3) migrate from injured sites to regional lymph nodes and; (4) sensitize naive T cells *in situ*. Although several tissue-derived cells have been found to be capable of presenting antigen to memory-type T cells, including dermal fibroblasts, endothelial cells and melanocytes,[31,32] the sensitization of naive T cells has been considered to be a particular function of dendritic cells.[25,26] Fibrocytes are distinct from dendritic cells and their precursors not only in their growth properties (fibrocytes are an adherent, proliferating cell population whereas dendritic cells are non-adhering and poorly proliferating) but also in their surface protein expression ($collagen^+/CD13^+/CD34^+/CD25^-/CD10^-/CD38^-$). That fibrocytes also have this specialized and potent priming activity suggests that they may play a critical role in the initiation of immunity during tissue injury and repair.

In mice, ~5% of fibrocytes were found to home to regional lymph nodes after intradermal injection into the skin. Fibrocytes thus may function *in vivo* to capture foreign proteins at the sites of tissue injury and to migrate into regional lymph nodes for the purpose of sensitizing naive T cells and/or activating memory T cells. Once naive T cells are sensitized by fibrocytes, they can migrate from draining lymph nodes via the circulation to sites of inflammation and perform effector functions.[33] Specifically, these sensitized T cells can act in an antigen-specific manner to induce B cells to proliferate and produce antibody as well as to induce macrophages to differentiate and produce proteins required for cell-mediated immunity (e.g. TNF-α). Peripheral blood fibrocytes are potent primers of naive T cells and thus may play an important role in the initiation of this cascading series of cellular events required for immunity.

Certain professional APCs are essential for the induction of $CD8^+$ T cell-mediated immunity.[34,35] Dendritic cells bearing antigen are thought to sensitize resting $CD8^+$ T cells which then acquire the ability to kill infected targets. However, certain "non-professional" tissue fibroblasts expressing viral proteins are also capable of inducing a strong $CD8^+$ T cell response *in vivo* without any involvement of host dendritic cells. The role of peripheral blood fibrocytes in $CD8^+$ T cell-mediated immunity is thus of particular interest since, in addition to expressing several fibroblast-like features, they are capable of priming naive $CD4^+$ T cells *in situ*.

Recent studies by Balmelli *et al.* have demonstrated that fibrocytes isolated from porcine blood can activate $CD8^+$ cytolytic T lymphocytes (CTL) against classical swine fever virus (CSFV), an RNA virus that causes lethality in pigs.[36] Like mouse and human fibrocytes, porcine fibrocytes were found to express a sufficient complement of surface proteins required for antigen presentation, including class I and II MHC and the co-stimulatory accessory molecules, CD80 and CD86.[36] Additionally, porcine fibrocytes were capable of activating $CD4^+$ T lymphocytes as observed in fibrocytes isolated from humans and mice.[36] CSFV-infected fibrocytes were incubated with autologous lymphocytes from a CSFV-immune pig and found to markedly stimulate proliferation.[36] Saturating concentrations of anti-CD8 monoclonal antibody suppressed the proliferation, supporting the conclusion that fibrocytes can stimulate $CD8^+$ T cells.[36] As further evidence of the capacity of fibrocytes to activate $CD8^+$ T cells, CSFV-specific T lymphocytes were found to surround and kill CSFV-infected fibrocytes but not uninfected fibrocytes. Perhaps most interesting, co-cultures of CSFV-infected fibrocytes with immune lymphocytes at a low APC/T lymphocyte ratio (1:400) caused a large increase in $CD4^-CD8^{high}CD25^{high}$ cells, whereas co-cultures of CSFV-infected dendritic cells at a similar ratio had little effect.[36] Taken together, these data indicate that

$CD8^+$ T lymphocytes are able to specifically recognize and lyse virus-infected fibrocytes and thus demonstrate the capacity of fibrocytes to present exogenous antigen via class I MHC.

That peripheral blood fibrocytes can activate $CD8^+$ T lymphocytes is of particular interest since $CD8^+$ T lymphocytes can be primed to selectively kill neoplastic cells. In recent studies, mouse fibrocytes were pulsed with a protein lysate of D5 melanoma cells and then injected subcutaneously into the hind leg of C57Bl/6 mice twice over seven days.[37] The mice were then challenged with live D5 melanoma cells which grow as solid tumors. After 22 days, mice vaccinated with fibrocytes pulsed with the D5 lysate displayed significantly decreased growth compared with mice vaccinated with fibrocytes pulsed with an irrelevant lysate of MCA-207 sarcoma cells.[37] These data suggest that fibrocytes may play a role in tumor surveillance and thus may find clinical utility as a cell therapy against neoplastic growth.

The precise contribution of fibrocytes to antigen presentation processes during tissue repair will not be known until a method for specifically depleting or neutralizing these cells becomes available. As such, the future development of cytotoxic monoclonal antibodies which are specific for surface antigens only present on fibrocytes will greatly assist in the study of the functional role of this cell type *in vivo*. Fibrocyte-specific monoclonal antibodies will also assist in quantifying these cells in experimental lesions and provide an important tool to measure fibrocytes in peripheral blood. Additionally, the characterization of the cell differentiation pathway of fibrocytes may provide targets for the production of a transgenic fibrocyte "knock-out" mouse which could also prove useful in the assessment of their relative role in antigen presentation processes *in vivo*.

Fibrocytes Secrete Type I Collagen and Inflammatory Cytokines

Cytokine modulation of cellular function is essential for the host reparative response to local injury and for the induction of T cell-mediated immunity.[21] Peripheral blood leukocytes and other cell types produce a variety of inflammatory and fibrogenic cytokines that serve to coordinate successive cellular events in the tissue repair response.[38] Early inflammatory cytokines released into the wound micro-environment (e.g. MIP-1α and MIP-2) activate and attract blood-borne leukocytes and connective tissue cells.[39] Fibrogenic cytokines such as TGF-β1 and TNF-α stimulate connective tissue cells to proliferate and to secrete extracellular matrix proteins required for connective tissue scar formation.[40,41]

Studies employing *in situ* hybridization, RT-PCR and immunohistochemistry have established that macrophages are the primary source of inflammatory and fibrogenic cytokines during the host response to tissue injury. Macrophages are thus considered the master conductor of the tissue repair response, influencing inflammation, fibrosis and angiogenesis. Given that fibrocytes enter the sites of tissue injury rapidly, the secretory capacity of these cells has been intensely studied.

Mouse fibrocytes isolated from wound chambers express mRNAs for the proinflammatory cytokines, IL-1β and TNF-α, and the anti-inflammatory cytokine, IL-10.[42] Mouse fibrocytes also express high levels of MIP-1 and MIP-2, the fibrogenic growth factors, PDGF-A and TGF-β1, and the hemopoietic growth factor M-CSF when compared with wound chamber associated monocytes or T cells.[42] Purified human fibrocytes similarly secrete the chemokines, MIP-1, MIP-1β, MCP-1, and the β chemokines, IL-8 and GRO.[42] The hemopoietic growth factors, M-CSF and IL-6, regulate macrophage differentiation and lymphocyte proliferation, respectively, and are known to be released during the early phase of tissue repair.[43,44] Peripheral blood fibrocytes constitutively secrete substantial amounts of M-CSF and IL-6, suggesting a regulatory role in the cellular response to tissue injury and repair.[42]

The parasitic disease schistosomiasis is characterized by a fibrosing, granulomatous reaction directed against parasite eggs that become entrapped in the hepatic and pulmonary circulations.[45] Spindle-shaped CD34$^+$ cells have been observed to localize to areas of connective tissue matrix deposition, suggesting that fibrocytes contribute to the fibrotic pathology that occurs as a consequence of *S. japonicum* infection.[42] Similar spindle-shaped $CD34^+$ cells have been observed in the inflammatory infiltrates of human atherosclerotic plaques, supporting a role for fibrocytes in the progression of coronary artery disease (Fig. 1). Coupled with the observation that fibrocytes secrete type I collagen, we thus postulate that fibrocytes

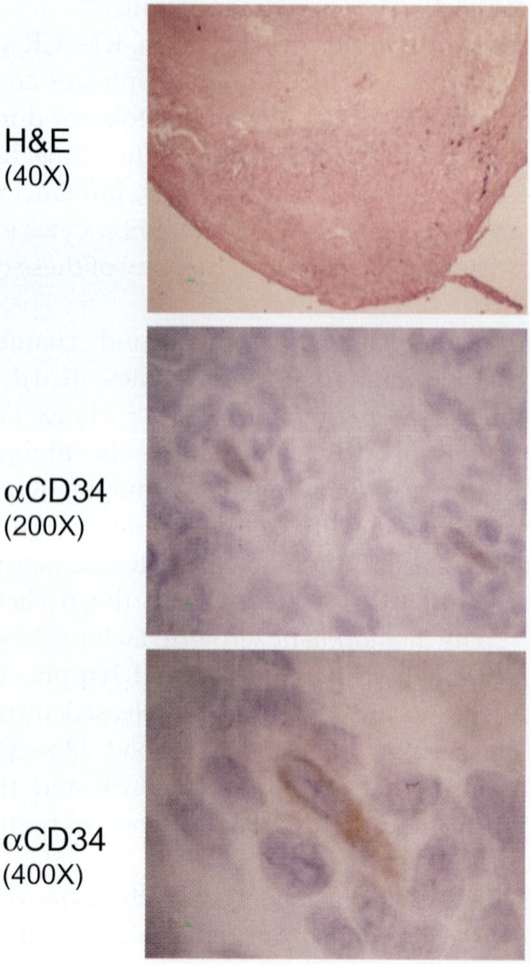

Fig. 1. Peripheral blood fibrocytes infiltrate human atherosclerotic plaques.

may serve a critical function in the fibrotic response to chronic inflammation.

In summary, fibrocytes purified from wound chambers implanted into mice and from human peripheral blood have been found to express chemokines, hematopoietic growth factors and fibrogenic cytokines. That fibrocytes are such an abundant source of essential cytokines suggests that these cells may play an important role in the control of both the inflammatory and repair phases of the wound healing response. The infiltration of $CD4^+$ T cells into areas of tissue damage is considered a critical requirement for the generation of antigen-specific immunity and the subsequent development of fibrosis. The fibrocyte products MIP-1α, MIP-1β and MCP-1 are potent T cell chemoattractants and may act to recruit $CD4^+$ T cells into the tissue repair micro-environment.[46,47] Fibrocytes may thus function to not only activate, but also recruit memory-type $CD4^+$ T cells.

Conclusion

The ability of fibrocytes to both recruit and activate T lymphocytes and to secrete type I collagen suggests that these cells may play a critical role in certain connective tissue disorders. A bidirectional fibrocyte: T lymphocyte activation response may lead to pathologically significant fibrosis since fibrocytes can amplify T lymphocyte infiltration and activation (Fig. 2). T lymphocytes are essential for the development of schistosomiasis and atherosclerosis and the localization of fibrocytes to areas of matrix deposition suggests that these cells may be an important source of collagen production in the schistosome-infected liver and atherosclerotic plaque. Fibrocytes may also participate in the generation of excessive fibroses in disorders that involve persistent T cell activation, including COPD, atherosclerosis, hepatitis, scleroderma, glomerulosclerosis, pulmonary fibrosis and graft versus host disease. Future studies which assess the type I collagen secretory response of fibrocytes upon interaction with T cells should provide important insight into the pathogenesis of these fibrotic disorders.

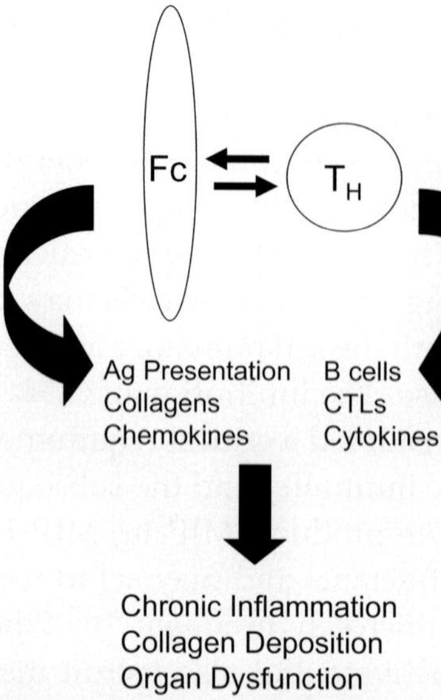

Fig. 2. Persistent bidirectional activation between fibrocytes and T lymphocytes as a mechanism of chronic fibrosis.

References

1. Wynn TA. (2004) Fibrotic disease and the T(H)1/T(H)2 paradigm. *Nat Rev Immunol* **4**: 583–594.
2. Hu H, Stein-Streilein J. (1993) Hapten-immune pulmonary interstitial fibrosis (HIPIF) in mice requires both $CD4^+$ and $CD8^+$ T lymphocytes. *J Leukoc Biol* **54**: 414–422.
3. Barbarin V, Arras M, Misson P, *et al.* (2004) Characterization of the effect of interleukin-10 on silica-induced lung fibrosis in mice. *Am J Respir Cell Mol Biol* **31**: 78–85.
4. Doege C, Koch M, Heratizadeh A, *et al.* (2005) Chronic allograft nephropathy in athymic nude rats after adoptive transfer of primed T lymphocytes. *Transpl Int* **18**: 981–991.
5. Suzuki N, Ohta K, Horiuchi T, *et al.* (1996) T lymphocytes and silica-induced pulmonary inflammation and fibrosis in mice. *Thorax* **51**: 1036–1042.
6. Schrier DJ, Kunkel RG, Phan SH. (1983) The role of strain variation in murine bleomycin-induced pulmonary fibrosis. *Am Rev Respir Dis* **127**: 63–66.

7. Szapiel SV, Elson NA, Fulmer JD, *et al.* (1979) Bleomycin-induced interstitial pulmonary disease in the nude, athymic mouse. *Am Rev Respir Dis* **120**: 893–899.
8. Wahl SM, Hunt DA, Allen JB, *et al.* (1986) Bacterial cell wall-induced hepatic granulomas. An *in vivo* model of T cell-dependent fibrosis. *J Exp Med* **163**: 884–902.
9. Vaage J. (1992) Immunologic aspects of fibrosis in mouse mammary carcinomas. *Int J Cancer* **50**: 69–74.
10. Faulkner MA, Hilleman DE. (2002) The economic impact of chronic obstructive pulmonary disease. *Expert Opin Pharmacother* **3**: 219–228.
11. Saetta M, Baraldo S, Corbino L, *et al.* (1999) $CD8^+$ve cells in the lungs of smokers with chronic obstructive pulmonary disease. *Am J Respir Crit Care Med* **160**: 711–717.
12. O'Shaughnessy TC, Ansari TW, Barnes NC, Jeffery PK. (1997) Inflammation in bronchial biopsies of subjects with chronic bronchitis: inverse relationship of $CD8^+$ T lymphocytes with FEV1. *Am J Respir Crit Care Med* **155**: 852–857.
13. Jonasson L, Holm J, Skalli O, *et al.* (1986) Regional accumulations of T cells, macrophages, and smooth muscle cells in the human atherosclerotic plaque. *Arteriosclerosis* **6**: 131–138.
14. Jonasson L, Holm J, Skalli O, *et al.* (1985) Expression of class II transplantation antigen on vascular smooth muscle cells in human atherosclerosis. *J Clin Invest* **76**: 125–131.
15. Stemme S, Faber B, Holm J, *et al.* (1995) T lymphocytes from human atherosclerotic plaques recognize oxidized low density lipoprotein. *Proc Natl Acad Sci USA* **92**: 3893–3897.
16. Plump AS, Smith JD, Hayek T, *et al.* (1992) Severe hypercholesterolemia and atherosclerosis in apolipoprotein E-deficient mice created by homologous recombination in ES cells. *Cell* **71**: 343–353.
17. Zhou X, Nicoletti A, Elhage R, Hansson GK. (2000) Transfer of $CD4^+$ T cells aggravates atherosclerosis in immunodeficient apolipoprotein E knockout mice. *Circulation* **102**: 2919–2922.
18. Daniil Z, Kitsanta P, Kapotsis G, *et al.* (2005) $CD8^+$ T lymphocytes in lung tissue from patients with idiopathic pulmonary fibrosis. *Respir Res* **6**: 81.
19. Abo-Zenah H, Katsoudas S, Wild G, *et al.* (2002) Early human renal allograft fibrosis: Cellular mediators. *Nephron* **91**: 112–119.

20. Koziel MJ. (1999) Cytokines in viral hepatitis. *Semin Liver Dis* **19**: 157–169.
21. Kovacs EJ. (1991) Fibrogenic cytokines: the role of immune mediators in the development of scar tissue. *Immunol Today* **12**: 17–23.
22. Kovacs EJ, DiPietro LA. (1994) Fibrogenic cytokines and connective tissue production. *Faseb J* **8**: 854–861.
23. Bucala R, Spiegel LA, Chesney J, *et al.* (1994) Circulating fibrocytes define a new leukocyte subpopulation that mediates tissue repair. *Mol Med* **1**: 71–81.
24. Petrakis NL, Davis M, Lucia SP. (1961) The *in vivo* differentiation of human leukocytes into histiocytes, fibroblasts and fat cells in subcutaneous diffusion chambers. *Blood* **17**: 109–118.
25. Inaba K, Metlay JP, Crowley MT, Steinman RM. (1990) Dendritic cells pulsed with protein antigens *in vitro* can prime antigen-specific, MHC-restricted T cells *in situ*. *J Exp Med* **172**: 631–640.
26. Levin D, Constant S, Pasqualini T, *et al.* (1993) Role of dendritic cells in the priming of $CD4^+$ T lymphocytes to peptide antigen *in vivo*. *J Immunol* **151**: 6742–6750.
27. Cassell DJ, Schwartz RH. (1994) A quantitative analysis of antigen-presenting **cell** function: activated B cells stimulate naive CD4 T cells but are inferior to dendritic cells in providing costimulation. *J Exp Med* **180**: 1829–1840.
28. Kupiec-Weglinski JW, Austyn JM, Morris PJ. (1988) Migration patterns of dendritic cells in the mouse. Traffic from the blood, and T cell-dependent and — independent entry to lymphoid tissues. *J Exp Med* **167**: 632–645.
29. Chesney J, Bacher M, Bender A, Bucala R. (1997) The peripheral blood fibrocyte is a potent antigen-presenting cell capable of priming naive T cells *in situ*. *Proc Natl Acad Sci USA* **94**: 6307–6312.
30. Sprent J. (1978) Restricted helper function of F1 hybrid T cells positively selected to heterologous erythrocytes in irradiated parental strain mice. II. Evidence for restrictions affecting helper cell induction and T-B collaboration, both mapping to the K-end of the H-2 complex. *J Exp Med* **147**: 1159–1174.
31. Geppert TD, Lipsky PE. (1985) Antigen presentation by interferon-gamma-treated endothelial cells and fibroblasts: differential ability to function as antigen-presenting cells despite comparable Ia expression. *J Immunol* **135**: 3750–3762.

32. Le Poole IC, Mutis T, van den Wijngaard RM, *et al.* (1993) A novel, antigen-presenting function of melanocytes and its possible relationship to hypopigmentary disorders. *J Immunol* **151**: 7284–7292.
33. Kripke ML, Munn CG, Jeevan A, *et al.* (1990) Evidence that cutaneous antigen-presenting cells migrate to regional lymph nodes during contact sensitization. *J Immunol* **145**: 2833–2838.
34. Bender A, Bui LK, Feldman MA, *et al.* (1995) Inactivated influenza virus, when presented on dendritic cells, elicits human $CD8^+$ cytolytic T cell responses. *J Exp Med* **182**: 1663–1671.
35. Bhardwaj N, Bender A, Gonzalez N, *et al.* (1994) Influenza virus-infected dendritic cells stimulate strong proliferative and cytolytic responses from human $CD8^+$ T cells. *J Clin Invest* **94**: 797–807.
36. Balmelli C, Ruggli N, McCullough K, Summerfield A. (2005) Fibrocytes are potent stimulators of anti-virus cytotoxic T cells. *J Leukoc Biol* **77**: 923–933.
37. Chesney J, Naftzger C, Rice G, *et al.* (2000) Vaccination with tumor lysate-pulsed peripheral blood fibrocytes provides protective immunity against the *in situ* growth of a murine melanoma. *Cellular Immunity Immunother Cancer* **110**: 66.
38. Thornton SC, Por SB, Walsh BJ, *et al.* (1990) Interaction of immune and connective tissue cells: I. The effect of lymphokines and monokines on fibroblast growth. *J Leukoc Biol* **47**: 312–320.
39. Oppenheim JJ, Zachariae CO, Mukaida N, Matsushima K. (1991) Properties of the novel proinflammatory supergene "intercrine" cytokine family. *Annu Rev Immunol* **9**: 617–648.
40. Battegay EJ, Raines EW, Seifert RA, *et al.* (1990) TGF-beta induces bimodal proliferation of connective tissue cells via complex control of an autocrine PDGF loop. *Cell* **63**: 515–524.
41. Leibovich SJ, Polverini PJ, Shepard HM, *et al.* (1987) Macrophage-induced angiogenesis is mediated by tumour necrosis factor-alpha. *Nature* **329**: 630–632.
42. Chesney J, Metz C, Stavitsky AB, *et al.* (1998) Regulated production of type I collagen and inflammatory cytokines by peripheral blood fibrocytes. *J Immunol* **160**: 419–425.
43. Paul WE, Seder RA. (1994) Lymphocyte responses and cytokines. *Cell* **76**: 241–251.
44. Sunderkotter C, Steinbrink K, Goebeler M, *et al.* (1994) Macrophages and angiogenesis. *J Leukoc Biol* **55**: 410–422.

45. Warren KS, Siongok TK, Houser HB, *et al.* (1978) Quantification of infection with Schistosoma haematobium in relation to epidemiology and selective population chemotherapy. I. Minimal number of daily egg counts in urine necessary to establish intensity of infection. *J Infect Dis* **138**: 849–855.
46. Loetscher P, Seitz M, Clark-Lewis I, *et al.* (1994) Monocyte chemotactic proteins MCP-1, MCP-2, and MCP-3 are major attractants for human $CD4^+$ and $CD8^+$ T lymphocytes. *Faseb J* **8**: 1055–1060.
47. Schall TJ, Bacon K, Camp RD, *et al.* (1993) Human macrophage inflammatory protein alpha (MIP-1α) and MIP-1β chemokines attract distinct populations of lymphocytes. *J Exp Med* **177**: 1821–1826.

Chapter 3

Regulatory Pathways for Fibrocyte Differentiation

Darrell Pilling* and Richard H. Gomer†

Introduction

Fibrosing diseases such as interstitial lung disease, cardiac fibrosis, keloid scarring, scleroderma, and severe asthma are chronic and debilitating conditions with a high mortality rate and in many western countries are an increasing problem. Interstitial lung disease is the general term for a broad category of lung diseases that includes idiopathic pulmonary fibrosis (IPF), a chronic and often fatal disorder which is characterized by fibrosis of the lungs. IPF has a survival rate of only 30% five years after diagnosis and an incidence of 1 in 400 in the elderly.[1] Cardiac fibrosis accounts for a significant fraction of the 450,000 deaths per year from cardiovascular disease in the US.[2,3]

*Department of Biochemistry and Cell Biology, MS-140, Rice University, Houston, TX 77005-1892, USA. Tel.: (+1) 713 348 4386; Fax: (+1) 713 348 5154. E-mail: dpilling@bioc.rice.edu
†Department of Biochemistry and Cell Biology, MS-140, Rice University, Houston, TX 77005-1892, USA. Tel.: (+1) 713 348 4872; Fax: (+1) 713 348 5154. E-mail: richard@bioc.rice.edu

Keloid and hypertrophic scarring affects 60–80% of burns patients and there are an estimated 11 million keloid-scarred individuals in the USA and Europe.[4,5] Scleroderma, although a relatively rare disease affecting 1 in 4,000 people, is increasing in frequency and is also associated with poor prognosis, with an average survival rate of 10 years.[6] Asthma affects approximately 10% of the population in the US, and 20% of asthma patients have chronic disease with fibrosis, which correlates with the severity of disease and poor prognosis.[7–11] There are currently no FDA approved therapeutics for fibrosis.[12]

Accelerated or aberrant fibrocyte differentiation appears to be at least in part responsible for fibrosing diseases.[13–16] Clearly, if we had a more complete understanding of how fibrocytes differentiate, the identity of factors that promote or inhibit fibrocyte differentiation, and the ability to regulate their numbers in fibrotic conditions, we may be able to design novel therapeutics to treat these diseases.

In order to describe more accurately the various populations of cells that we will discuss, we will first define these cells. **Monocytes** are CD14 positive peripheral blood mononuclear cells that enter tissues and have the ability to become a variety of cell types, including macrophages, fibrocytes, Langerhans' cells, and dendritic cells. **Fibrocyte precursors** are present in a subset (or multiple subsets) of monocytes that can become fibrocytes; these cells can be found in the bone marrow, the peripheral blood circulation, possibly the lymph, and in a tissue. **Early fibrocytes** are fibrocyte precursors that have committed to differentiation into fibrocytes but have not become elongated; we think that these events normally occur once the cell has entered a tissue. **Mature fibrocytes** are elongated fibroblast-like cells, and have been found in healing wounds, as well as in fibrotic lesions. Markers of mature fibrocytes include CD34, CD43, CD45, and collagens I and III. **Myofibrocytes** are derived from mature fibrocytes, and following activation with cytokines such as TGF-β, express alpha smooth muscle actin (α-SMA), and appear to have a similar function to myofibroblasts.

The factors that regulate fibrocyte differentiation are an active area of investigation and we expect to find that the list of regulators of fibrocyte differentiation will expand with time. However, there are probably four stages during which fibrocyte precursors and fibrocyte differentiation are likely to be controlled. These are: (1) In the bone marrow during the production of fibrocyte precursors; (2) in the peripheral circulation before the precursors enter tissues; (3) during the process of traversing the endothelium; and (4) in the tissues as fibrocytes differentiate, mature, and then become activated.

Biology of the Monocyte-Macrophage System

Peripheral blood monocytes (and by inference fibrocyte precursors) are part of the mononuclear phagocyte system (MPS) that develops from bone marrow precursors; and that differentiate into the various types of tissue macrophages.[17–19] Like all hematopoietic cells, the MPS are derived from a pluripotent stem cell, which gives rise to a common myeloid progenitor, which is thought to be responsible for all non-lymphoid lineages.[20] The common myeloid progenitors differentiate into either precursors of the megakaryocyte/erythroid lineage or the granulocyte/monocyte lineage (GM-CFU). The GM-CFU differentiates into either granulocytes (neutrophils) or monocytes. As with all hematopoietic cells, cells of the MPS are thought to follow the general paradigm that as differentiation progresses from the pluripotent stem cell into the defined cell type, their ability to differentiate into other cell types is lost or reduced.[21,22]

Peripheral blood monocytes are the intermediate, circulating precursor of tissue macrophages. Monocytes enter peripheral tissues and differentiate into a variety of macrophage lineage cells, including Kupffer cells of the liver, lung alveolar macrophages, brain microglia and peritoneal macrophages.[18] Under defined conditions, monocytes can also differentiate into a variety of other cells, including Langerhans' and dendritic cells, bone osteoclasts, adipocytes, and fibrocytes.[18,23–25] How the commitment to any particular end-stage cell is controlled and whether monocytes are a single population that alters its characteristics in response to specific microenvironments, or whether there are subsets of monocytes with different differentiation capabilities, is still under investigation.[26–29]

Differentiation of Monocytes into Cell Types Other than Macrophages

Dendritic cells (DCs) are the primary antigen-presenting cell responsible for stimulating naïve T cells and generating an adaptive immune response.[30,31] Like all other hematopoietic cells, DCs are continuously generated from stem cells in the bone marrow, and the mature DCs are present in all tissues of the body.[32,33] Immature, peripheral dendritic cells phagocytose either foreign or self-antigens, and then migrate to the draining lymph nodes where they differentiate into mature DCs and activate or tolerize T cells.[32,34] Dendritic cell lineages are complex, with DCs differentiating from both common lymphoid progenitors and common myeloid progenitors. How the differentiation of dendritic cells is regulated by exogenous factors, especially those related to infection and inflammation, is unclear. In humans, DCs are easily generated *in vitro* from peripheral blood monocytes, as well as from bone marrow and umbilical cord blood monocytes. The two main systems used to generate human myeloid DCs are either incubation with IL-4 and GM-CSF and subsequent maturation with TNF-α or bacterial products, such as lipopolysaccharide (LPS).[24,35] The second mechanism involves the migration of monocytes through an endothelial layer into a collagen matrix.[36,37] The cells that subsequently "transmigrate" back across the endothelial layer become DCs. The transmigration process mimics the tissue homing and trafficking of monocytes through the afferent lymph into draining lymph nodes.[38]

The factors that regulate the initial monocyte to DC differentiation involve positive factors such as GM-CSF plus IL-4, Flt3-Ligand, TNF-α, IL-15, transmigration, and CD40.[24,35,36,39,40] Factors that inhibit the initial differentiation of monocytes into DCs include IL-6, IL-10, IFN-γ, and bacterial products such as LPS.[41–44] These events are different from the known ability of bacterial, yeast, viral and helminth products to influence the maturation of DCs from immature dendritic cells.[34] These findings indicate that the environment encountered following migration into tissues can "fine tune" the final mature cell type.

Peripheral blood monocytes not only have the ability to become macrophages and dendritic cells but also a variety of other cell types. The osteoclast differentiates from a myeloid/monocyte lineage, and both bone marrow precursors and peripheral blood monocytes can become osteoclasts.[45,46] This process is dependent on RANKL and M-CSF and is enhanced by TGF-β.[47,48] Other cell types reported to be derived *in vitro* from monocytes include adipocytes as well as epithelial, endothelial, and neuronal cells.[49–51]

Peripheral Blood Mononuclear Cells can also Differentiate into Fibroblast/Stromal Cells

Fibroblasts repair wounds but also cause the hyperplasia characteristic of chronic inflammation.[52,53] There are two sources of these fibroblasts. First, local quiescent fibroblasts migrate into the affected area, producing extracellular matrix proteins, and promoting wound contraction.[54] In addition to the migration of local fibroblasts, circulating fibroblast precursors present within the blood are attracted to the sites of injury, where they differentiate into fibroblast-like cells (fibrocytes) and at least in part mediate tissue repair.[14,16,55] The suggestion that peripheral blood mononuclear cells can become connective tissue were made by Paget in 1863, Cohnheim in 1867, Metchnikoff in 1882, and Maximov in 1928.[15,16] Many subsequent studies in a variety of systems, including *in vitro* cultures of PBMCs, have confirmed this observation.[25,56–61] Richard Bucala and colleagues placed sponge-filled pieces of tubing in wounds in mice and observed that the tubing was quickly filled with neutrophils, monocytes, and leukocytes, but within 2 days of implantation, many of the cells in the tubes had a spindle shape and resembled fibroblasts.[59] To determine the origin of these fibroblast-like cells, these workers simply cultured human or mouse peripheral blood mononuclear cells (PBMCs) in a medium containing 20% heat-inactivated fetal calf serum and observed spindle-shaped cells in the culture after approximately 2 weeks. These spindle-shaped cells, which were named fibrocytes (fibrocyte is a mix of FIBROblast and lymphoCYTE), expressed markers of both hematopoietic (CD45, MHC class II,

CD34) and stromal cells (collagen I and III and fibronectin).[59,62] Fibrocytes are also potent antigen presenting cells capable of initiating a naïve T cell response, with similar kinetics to dendritic cells.[63,64]

In culture, human fibrocytes proliferate with a doubling time of 3 to 4 days.[59] Fibrocyte precursors appear to be a subpopulation of $CD14^+$ peripheral blood monocytes.[25,60,61] Together, the data suggest that blood monocytes can differentiate into fibroblast-like cells, and support the hypothesis that peripheral blood-derived cells are (at least in part) responsible for wound repair. A puzzling question that remains from this work is the discrepancy in the time course of fibrocyte differentiation — monocytes appear to differentiate into fibrocytes in 2 days in wound chambers, but take about 2 weeks to differentiate in culture. Our work suggests that serum amyloid P, a serum protein, inhibits fibrocyte differentiation, and that the serum present in the culture medium is one reason fibrocytes take longer to appear in cultures containing serum than in wound chambers.

Soluble Factors that Regulate Fibrocyte Differentiation

The control of fibrocyte differentiation is likely to occur at multiple points. However, the factors that regulate these events are still poorly understood. Several groups have shown that cytokines, especially "inflammatory" cytokines such as TNF-α, TGF-β, and IL-1β, have major roles in regulating **mature** fibrocyte activation (Fig. 1). IL-1β and TNF-α regulate fibrocytes in a similar but not identical manner.[25,62,64] TNF-α, and especially IL-1β, promote MHC class I expression and modest fibrocyte proliferation, both in humans and in pigs.[62,64] TNF-α and IL-1β also promote chemokine production from mature fibrocytes, with increased levels of CCL2 (MCP-1), CCL3 (MIP-1α), and CCL4 (MIP-1β), which are all likely to act on activated T cells, immature dendritic cells, NK cells and monocytes. TNF-α and IL-1β also induce mature fibrocytes to secrete chemokines that will act on neutrophils, such as CXCL1 (GRO-α) and CXCL8 (IL-8). IL-1β also generates a feedback loop to increase IL-10 and TNF-α secretion by mature fibrocytes, and reduces collagen-I production. Therefore,

IL-1β produced by monocyte/macrophages, T cells, and many other inflammatory cells, but not fibrocytes, may be a mechanism to prevent mature fibrocytes from differentiating into myofibrocytes.

TGF-β promotes the differentiation of mature fibrocytes into α-SMA positive myofibrocytes.[25,60,65–67] These α-SMA positive fibrocytes are able to contract collagen gels and produce increased amounts of collagen-I and fibronectin, suggesting that they are the cells responsible for wound healing. These cells have also been observed *in vivo* by several groups (see others chapters in this book for details). Although the IL-1β, TNF-α, and TGF-β signal transduction pathways have been elucidated for other cell types, the corresponding fibrocyte signal transduction pathways have not been described.

To identify signals that affect the **initial** differentiation of fibrocytes, we culture PBMC in a serum-free medium system to reduce

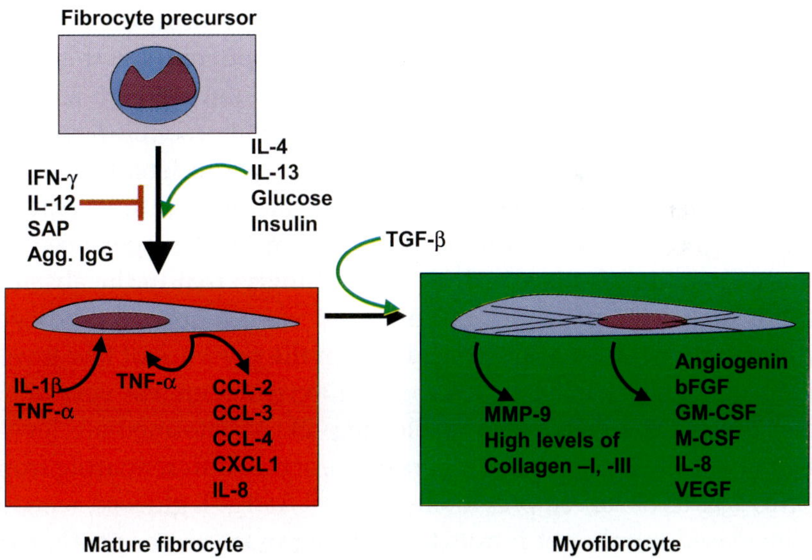

Fig. 1. Regulation of fibrocyte differentiation. Fibrocyte differentiation in the peripheral tissues is regulated at two key stages. The first is during differentiation from the circulating fibrocyte precursor cells into mature fibrocytes. The second is during transformation into myofibrocytes.

any unwanted interactions between the fibrocyte precursors and possible ligands present in serum, such as cytokines, growth factors, IgG, or serum amyloid P (SAP) (Fig. 1). We found that pro-inflammatory cytokines such as IL-1β, IL-6, TNF-α, IL-16, and IL-18 had no effect on the numbers of fibrocytes differentiating in serum-free cultures of PBMC. However, the pro-inflammatory cytokines IL-12 and IFN-γ inhibited fibrocyte differentiation, and the profibrotic cytokines IL-4 and IL-13 promoted fibrocyte differentiation (manuscript in preparation). In mixing experiments we found that IL-4 had a dominant effect over IL-12 or IFN-γ. These data suggest that the cytokine milieu present in a tissue will have a profound effect on the number of fibrocyte precursors differentiating into mature fibrocytes.

Regulation of Fibrocytes by Glucose and Insulin

Since the original description of fibrocytes in the pva-sponge wound model, fibrocytes have been observed in many healing and non-healing wounds.[25,59,60,62,65–73] The authors wondered whether conditions that are well known to have poor wound healing, such as those in diabetic patients, were due to an inhibition of fibrocyte differentiation or an inability to support fibrocyte differentiation. To test this, we cultured PBMC in serum-free medium with increasing levels of glucose with insulin at 10 μg/ml, or increasing concentrations of insulin. We observed (not surprisingly) that in the absence of either glucose or insulin not only did we not observe fibrocyte differentiation, but that the cells died!! In the presence of low levels of glucose (0.004–0.025 g/L), fibrocyte differentiation was very poor, even though the cells were viable (Fig. 2). At intermediate levels of glucose (0.1–2 g/L) the numbers of fibrocytes differentiating from PBMC was as observed previously.[25,61,65,74] At 8 g/L glucose, we saw an increased number of fibrocytes. Adding pyruvate, as an alternative carbon source, also augmented the number of fibrocytes (Fig. 2). Increasing the concentration of insulin in the cell culture conditions also promoted fibrocyte differentiation, especially at higher glucose concentrations. These data suggest that *in vitro* analysis of fibrocytes

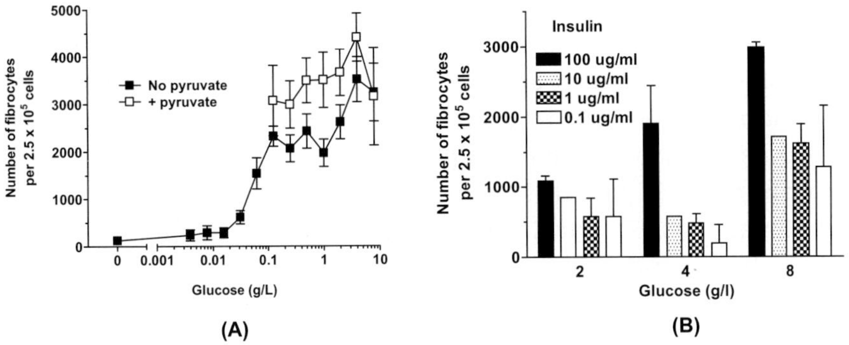

Fig. 2. Regulation of fibrocyte differentiation by glucose and insulin. (**A**) Human PBMCs at 2.5×10^5 per ml were cultured in serum-free medium containing $10\,\mu g/ml$ insulin for 5 days in the presence ($n = 3$) or absence ($n = 7$) of 1 mM sodium pyruvate and increasing concentrations of glucose. Cells were then air-dried, fixed, stained and fibrocytes enumerated by morphology. (**B**) PBMCs were cultured in 2, 4 or 8 g/l glucose with increasing concentrations of insulin. Results are mean ± SD of fibrocytes per 2.5×10^5 PBMCs.

using defined media such as RPMI-1640 (2 g/L glucose) or DMEM (4.5 g/L glucose), without adding supplements will lead to suboptimal differentiation of fibrocytes. The addition of 10% serum (assuming serum contains between 1–2 g/L glucose) will only add 0.2 g/L to the system.

Regulation of Fibrocyte Differentiation by SAP and Aggregated IgG

While examining the possible role of cell density in the survival of peripheral blood T cells,[75–77] we observed that in serum-free medium, PBMCs gave rise to a population of fibroblast-like cells within 3 days.[61] These cells were adherent and had a spindle-shaped morphology. When plasma or serum was present at concentrations between 10% and 0.1%, there was a significant reduction in the number of the fibroblast-like cells. However, at or below 0.01% plasma or serum, fibroblast-like cells rapidly developed. The activity in the serum or plasma that inhibited fibrocyte formation was retained by a

30 kDa cut-off spin-filter. If the serum was heated to 56°C for 30 min, the efficacy was reduced approximately 10-fold, and heating to 95°C abolished the inhibitory activity. The inhibitory factor appeared to be evolutionary conserved, as bovine, goat, horse, murine and rat sera were also able to inhibit the appearance of these fibroblast-like cells from human peripheral blood precursors. In serum-free culture, ~1% of human, rat and murine PBMCs differentiate into spindle-shaped cells. Human, rat and mouse sera also inhibit the differentiation of fibrocytes from rat PBMCs, and human and murine sera inhibit the differentiation of murine PBMCs into fibrocytes.

Using conventional protein biochemistry, we were able to purify the activity in human serum that inhibits fibrocyte differentiation and identify it as serum amyloid P (SAP).[61] A commercial preparation of SAP was able to inhibit fibrocyte differentiation, whereas other serum proteins such as protein S, serum amyloid A (SAA), and the highly related protein C-reactive protein (CRP) could not (Fig. 3).[61,74] To confirm that SAP is the active factor in a serum that inhibits fibrocyte differentiation, we depleted SAP using anti-SAP antibodies bound to protein G beads, or by calcium-dependent binding of SAP to high EEO agarose. The SAP-depleted serum had a poor ability to inhibit fibrocyte differentiation.[61] Together with the ability of purified SAP to inhibit fibrocyte differentiation, these observations strongly suggested that SAP is the active factor in serum and plasma that inhibit fibrocyte differentiation.

SAP, a member of the pentraxin family of proteins that includes C-reactive protein (CRP), is produced by the liver, secreted into the blood, and which circulates in the blood as stable pentamers.[78–81] It is important to note that that serum amyloid P (SAP) is **not** serum amyloid A, serum amyloid protein, or amyloid precursor protein.[82] SAP appears to play a role in both the initiation and resolution phases of the immune response.[83–85] SAP binds to sugar residues on the surface of bacteria, leading to their opsonization and engulfment.[81,83] SAP also binds to free DNA and chromatin generated by apoptotic cells at the resolution of an immune response, thus preventing a secondary inflammatory response.[84] Bacteria and proteins bound by SAP are removed by phagocytic cells, such as macrophages, due to

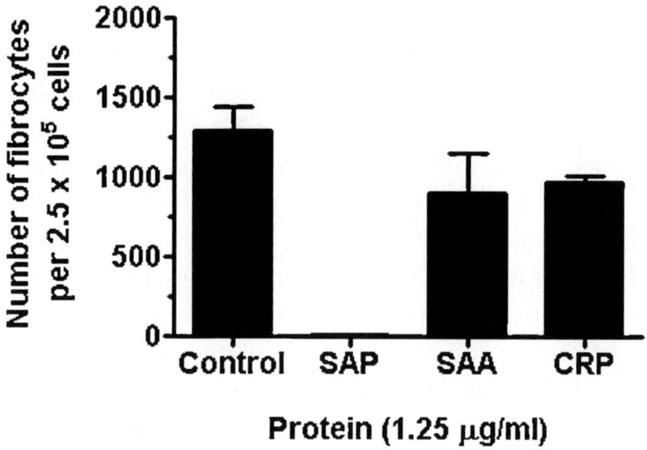

Fig. 3. Inhibition of fibrocyte differentiation by SAP but not CRP or other plasma proteins. Human PBMCs at 2.5×10^5 per ml were cultured in serum-free medium for 5 days in the presence of commercially available purified SAA, SAP, or CRP and were then examined for the appearance of fibroblast-like cells. Cells were air-dried, fixed, stained and fibrocytes enumerated by morphology. Results are mean ± SEM of fibrocytes per 2.5×10^5 PBMCs ($n = 3$ separate individuals). SAA at 1.25 μg/ml (100 nM) had no effect on fibrocyte differentiation, whereas 1.25 μg/ml SAP (10 nM) completely inhibited fibrocyte differentiation. When this assay was repeated using PBMCs from other donors, similar results were obtained with the exception that the number of fibrocytes in serum-free medium varied from 1,500 to 4,000 per 2.5 × 105 cells, depending on the donors.

the ability of SAP to bind to all three classical FcγR, with a preference for FcγRI and FcγRII.[86,87] CRP appears to bind with a high affinity to FcγRII, a lower affinity to FcγRI, but does not bind FcγRIII.[88–92] Both SAP and CRP initiate intracellular signaling events consistent with FcγR ligation.[86,89,92,93]

As SAP inhibits fibrocyte differentiation and binds to FcγRs, and IgG also binds FcγRs, we added IgG to the peripheral blood mononuclear cells in a serum-free medium to see if IgG could also inhibit fibrocyte differentiation. We found that aggregated IgG could inhibit human fibrocyte differentiation, but there was no inhibition by monomeric IgG, IgA, IgE, or IgM, or aggregated IgA, IgE, or IgM.[74] Monoclonal $F(ab')_2$ antibodies specific for FcγRI, and FcγRII, when cross-linked, also inhibited fibrocyte differentiation. Consistent with the hypothesis that cross-linked IgG inhibits fibrocyte differentiation by multimerizing FcγRs, removal of the Fc region from IgG abrogated the ability of IgG to inhibit fibrocyte differentiation, and pharmacological inhibition of the FcγR signal transduction pathway in monocytes inhibited the ability of SAP or cross-linked IgG to inhibit fibrocyte differentiation (Figs. 4 and 5).

The mechanisms by which SAP and aggregated IgG inhibit fibrocyte differentiation appear to be distinct but related. The inhibition of fibrocyte differentiation by SAP was dependent on src-related tyrosine kinases (SRTK), as this inhibition was PP2 sensitive. By comparison, aggregated IgG inhibited fibrocyte differentiation by a SRTK and Syk-dependent process. These data suggest that both SAP and IgG activate one or more SRTK, but the subsequent events are divergent. Alternatively, the differences observed could be explained if SAP preferentially binds FcγRII, since FcγRII signaling is more resistant to Syk antagonists.[94–96]

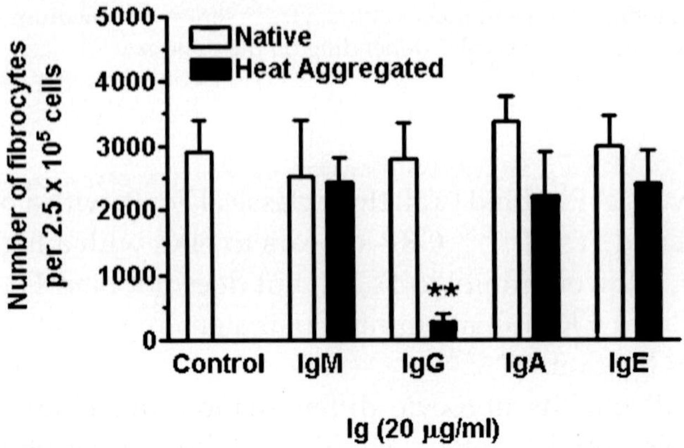

Fig. 4. Cross-linked but not monomeric IgG inhibits fibrocyte differentiation. PBMCs at 2.5×10^5 per ml were cultured in serum-free medium for 5 days in the presence or absence of 20 μg/ml native or heat-aggregated IgM, IgG, IgA or IgE. Only aggregated IgG inhibited fibrocyte differentiation. Values are means $\pm$ SEM ($n = 4$). $^{**}p < 0.01$. Figure reproduced from Ref. 74, with permission from the Society for Leukocyte Biology.

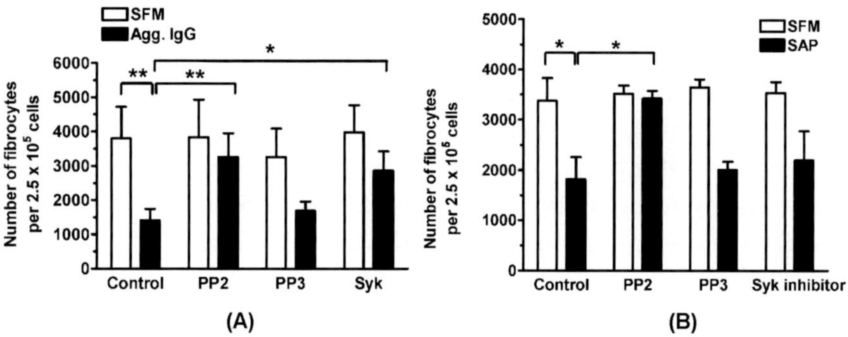

Fig. 5. Direct ligation of FcγR on monocytes leads to the inhibition of fibrocyte differentiation. PBMC at 2.5×10^5 per ml were incubated for 60 min at 37°C, and non-adherent cells were then removed. (**A**) The adherent monocytes were incubated for 60 min at 4°C in the presence or absence of 10 nM PP2, PP3 or Syk inhibitor. Monocytes were then washed twice, and cultured in the presence or absence of heat-aggregated human IgG for 60 min at 4°C. Monocytes were then washed twice, and the non-adherent cells replaced to a final concentration of 2.5×10^5 cells/ml and then cultured for 5 days at 37°C in serum-free medium. Compared to monocytes cultured in serum-free medium, aggregated IgG significantly inhibited fibrocyte differentiation ($p < 0.01$), as determined by ANOVA. Compared to monocytes incubated with 10 μg/ml aggregated IgG, pre-incubation with PP2 ($p < 0.01$; **) or the Syk inhibitor ($p < 0.05$; *) significantly inhibited the ability of IgG to inhibit fibrocyte differentiation as determined by ANOVA. (**B**) Adherent monocytes were incubated for 60 min at 4°C in the presence or absence of 10 nM PP2, PP3 or Syk inhibitor. Monocytes were then washed and incubated for 60 min at 4°C in the presence or absence of 0.5 μg/ml SAP. Cells were then washed, the non-adherent cells were replaced and the cells were cultured for 5 days. Compared to monocytes cultured in SFM, SAP significantly inhibited fibrocyte differentiation ($p < 0.05$), as determined by ANOVA. Compared to monocytes cultured in 0.5 μg/ml SAP, pre-incubation with PP2 significantly inhibited the ability of SAP to prevent fibrocyte differentiation ($p < 0.05$; *), as determined by ANOVA. Results are expressed as the mean ± SEM of the number of fibrocytes per 2.5×10^5 cells ($n = 3$ separate donors). Figure reproduced from Ref. 74, with permission from the Society for Leukocyte Biology.

During an initial immune response, IgG molecules bound to pathogens form an immune complex that activate FcγR.[97,98] Our data suggest that these complexes would inhibit the differentiation of monocytes into fibrocytes. However, during the resolution phase, IgG and many other serum proteins such as SAP, are cleared from the site of infection by four main mechanisms: restoration of hemostasis following repair of blood vessels; drainage of the wound fluid through the lymphatic system; engulfment by phagocytic cells; and degradation by proteases.[99–102] Assuming that both SAP and immune complexes are cleared from the tissue during the resolution phase, the resulting low levels of SAP and immune complexes would create an environment favorable for the differentiation of fibrocytes, and thus aid tissue regeneration and wound healing.[53,103]

Regulation of Fibrocyte Differentiation by T cells and Extracellular Matrix

At least two other factors promote the differentiation of monocytes into fibrocytes. First, T cells are necessary for fibrocyte differentiation.[25,60] Whether this is a direct cell:cell interaction or a soluble molecule released by co-culturing T cells and monocytes together is still unclear, and the nature of this profibrocyte molecule is as yet unknown. In preliminary experiments, we have also found that conditioned medium from T cells can promote fibrocyte differentiation from a purified CD14-positive population. Using blocking anti-cytokine antibodies have excluded the two profibrocyte cytokines IL-4 and IL-13, and currently attempts are made to identify the molecules involved.

We have also found that extracellular matrix (ECM) proteins have an inhibitory effect on fibrocyte differentiation. If PBMC are cultured in a serum-free medium in the presence of the ECM proteins, collagen-I (Col I) or Pronectin-F (a synthetic RGD-motif from fibronectin (Fn)) fibrocyte differentiation was inhibited (Fig. 6). During the initial response to trauma or infection when tissue injury and damage to the vascular architecture occurs, fibrocyte precursors may encounter these ECM proteins, either in isolation or as degraded proteins. This may then impart an inhibitory signal to the fibrocyte precursors, resulting in monocyte to macrophage differentiation. However, once blood vessels have been repaired, and the basement membrane has been reformed, then fibrocyte precursors passing through the vessel wall may receive signals to promote fibrocyte differentiation. The ability of SAP and ECM proteins (or ECM

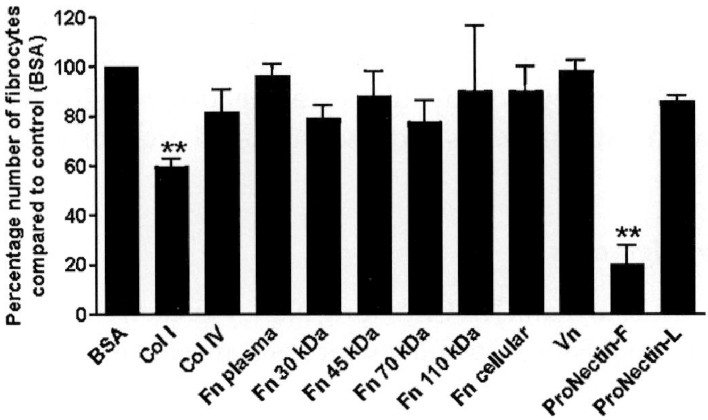

Fig. 6. Regulation of fibrocyte differentiation by extracellular matrix proteins. Human PBMCs at 2.5×10^5 per ml were cultured in serum-free medium for 5 days in the presence of bovine serum albumin (BSA; control) or ECM proteins (10 μg/ml) except vitronectin (Vn 1 μg/ml). Cells were air-dried, fixed, stained and fibrocytes enumerated by morphology ($n = 4$ individuals). Collagen-I (Col I) and ProNectin-F significantly inhibited fibrocyte differentiation, as determined by ANOVA (** = $p < 0.01$).

fragments) to inhibit fibrocyte differentiation suggests that the presence of SAP or ECM proteins in the peripheral tissues would be a means of regulating monocyte to fibrocyte differentiation. Interestingly, in peripheral tissues, SAP is associated with several ECM proteins, such as elastin in the skin and collagen and laminin in the basement membrane.[104–109]

Summary

Fibrocytes have emerged as a key cell in fibrosing diseases. There is currently no FDA-approved therapy for these diseases. Inhibiting the appearance of fibrocytes in a tissue could inhibit the progress of a fibrosis. Work from several laboratories has indicated that there are multiple steps in the fibrocyte differentiation pathway that are regulated either in a positive or a negative manner by secreted factors or other elements of the tissue environment. Controlling or manipulating this regulation may allow us to eventually prevent fibrocytes from appearing in a fibrotic lesion, and thus at least stop the progression of the disease. Finally, finding factors that are required to maintain the fibrocyte or myofibrocyte phenotype might eventually allow us to manipulate this and reverse the progression of a fibrosis.

Acknowledgments

We would like to thank Jeff Crawford, Nancy Tucker, Hillary Patuwo, Diane Shao and Sanna Ronkainen for assistance. This work was supported by grant number C-1555 from the Robert A. Welch Foundation and grant 005/04 from the Scleroderma Foundation.

References

1. Perez A, Rogers RM, Dauber JH. (2003) The prognosis of idiopathic pulmonary fibrosis. *Am J Resp Cell Mole Bio* **29**: S19–S26.
2. Assayag P, Carre F, Chevalier B, *et al.* (1997) Compensated cardiac hypertrophy: arrhythmogenicity and the new myocardial phenotype. I. fibrosis. *Cardiovasc Res* **34**: 439–444.
3. Haider AW, Larson MG, Benjamin EJ, Levy D. (1998) Increased left ventricular mass and hypertrophy are associated with increased risk for sudden death. *J Am Coll Cardiol* **32**: 1454–1459.
4. Bombaro KM, Engrav LH, Carrougher GJ. (2003) What is the prevalence of hypertrophic scarring following burns? *Burns* **29**: 299–302.
5. Bayat A, McGrouther DA, Ferguson MW. (2003) Skin scarring. *BMJ* **326**: 88–92.
6. Mayes MD, Lacey JV Jr, Beebe-Dimmer J. (2003) Prevalence, incidence, survival, and disease characteristics of systemic sclerosis in a large US population. *Arthritis Rheum* **48**: 2246–2255.
7. Roche WR, Beasley R, Williams JH, Holgate ST. (1989) Subepithelial fibrosis in the bronchi of asthmatics. *Lancet* **1**: 520–524.
8. Chetta A, Foresi A, Del Donno M, *et al.* (1997) Airways remodeling is a distinctive feature of asthma and is related to severity of disease. *Chest* **111**: 852–857.
9. Brewster CE, Howarth PH, Djukanovic R, *et al.* (1990) Myofibroblasts and subepithelial fibrosis in bronchial asthma. *Am J Respir Cell Mol Biol* **3**: 507–511.

10. Davies DE, Wicks J, Powell RM, *et al.* (2003) Airway remodeling in asthma: new insights. *J Allergy Clin Immunol* **111**: 215–225.
11. Elias JA, Zhu Z, Chupp G, Homer RJ. (1999) Airway remodeling in asthma. *J Clin Invest* **104**: 1001–1006.
12. Bouros D, Antoniou KM. (2005) Current and future therapeutic approaches in idiopathic pulmonary fibrosis. *Eur Respir J* **26**: 693–703.
13. Chesney J, Bucala R. (2000) Peripheral blood fibrocytes: mesenchymal precursor cells and the pathogenesis of fibrosis. *Curr Rheumatol Rep* **2**: 501–505.
14. Metz CN. (2003) Fibrocytes: a unique cell population implicated in wound healing. *Cell Mol Life Sci* **60**: 1342–1350.
15. Postlethwaite AE, Shigemitsu H, Kanangat S. (2004) Cellular origins of fibroblasts: possible implications for organ fibrosis in systemic sclerosis. *Curr Opin Rheumatol* **16**: 733–738.
16. Quan TE, Cowper S, Wu S-P, *et al.* (2004) Circulating fibrocytes: collagen-secreting cells of the peripheral blood. *Int J Biochem Cell Biol* **36**: 598–606.
17. Van Furth R, Cohn ZA, Hirsch JG, *et al.* (1972) The mononuclear phagocyte system: a new classification of macrophages, monocytes, and their precursor cells. *Bull World Health Organ* **46**: 845–852.
18. Gordon S. (1999) Development and distribution of mononuclear phagocytes: Relevance to Inflammation. In: Gallin JI, Synderman R (eds.), *Inflammation: Basic Principles and Clinical Correlates*, Lippincott, Williams and Wilkins, Philadelphia, pp. 35–48.
19. Sasomo RT, Hume DA. (2004) The biology of macrophages. In: Kaufmann SH, Medzhitov R, Gordon S (eds.), *The Inate Immune Response to Infection*, ASM Press, Washington D.C., pp. 71–93.
20. Ross JA, Auger MJ. (2002.) The biology of the macrophage. In: Burke B, Lewis CE (eds.), *The Macrophage*, Oxford University Press, Oxford, pp. 1–72.
21. Goodell MA. (1999) $CD34^+$ or $CD34^-$: does it really matter? *Blood* **94**: 2545–2547.
22. Shostak S. (2006) (Re)defining stem cells. *Bioessays* **28**: 301–308.
23. Inaba K, Inaba M, Romani N, *et al.* (1992) Generation of large numbers of dendritic cells from mouse bone-marrow cultures supplemented with granulocyte macrophage colony-stimulating factor. *J Exp Med* **176**: 1693–1702.
24. Sallusto F, Lanzavecchia A. (1994) Efficient presentation of soluble antigen by cultured human dendritic cells is maintained by

granulocyte/macrophage colony-stimulating factor plus interleukin 4 and downregulated by tumor necrosis factor alpha. *J Exp Med* **179**: 1109–1118.

25. Abe R, Donnelly SC, Peng T, *et al.* (2001) Peripheral blood fibrocytes: differentiation pathway and migration to wound sites. *J Immunol* **166**: 7556–7562.
26. Passlick B, Flieger D, Ziegler-Heitbrock HW. (1989) Identification and characterization of a novel monocyte subpopulation in human peripheral blood. *Blood* **74**: 2527–2534.
27. Grage-Griebenow E, Flad HD, Ernst M. (1996) Fc gamma receptor I (CD64)-negative human monocytes are potent accessory cells in viral antigen-induced T cell activation and exhibit high IFN-alpha-producing capacity. *J Leukoc Biol* **60**: 389–396.
28. Geissmann F, Jung S, Littman DR. (2003) Blood monocytes consist of two principal subsets with distinct migratory properties. *Immunity* **19**: 71–82.
29. Sunderkotter C, Nikolic T, Dillon MJ, *et al.* (2004) Subpopulations of mouse blood monocytes differ in maturation stage and inflammatory response. *J Immunol* **172**: 4410–4417.
30. Steinman RM. (1991) The dendritic cell system and its role in immunogenicity. *Annu Rev Immunol* **9**: 271–296.
31. Shortman K, Liu YJ. (2002) Mouse and human dendritic cell subtypes. *Nat Rev Immunol* **2**: 151–161.
32. Liu YJ. (2001) Dendritic cell subsets and lineages, and their functions in innate and adaptive immunity. *Cell* **106**: 259–262.
33. Manz MG, Traver D, Miyamoto T, *et al.* (2001) Dendritic cell potentials of early lymphoid and myeloid progenitors. *Blood* **97**: 3333–3341.
34. Mellman I, Steinman RM. (2001) Dendritic cells: specialized and regulated antigen processing machines. *Cell* **106**: 255–258.
35. Chomarat P, Dantin C, Bennett L, *et al.* (2003) TNF skews monocyte differentiation from macrophages to dendritic cells. *J Immunol* **171**: 2262–2269.
36. Randolph GJ, Beaulieu S, Lebecque S, *et al.* (1998) Differentiation of monocytes into dendritic cells in a model of transendothelial trafficking. *Science* **282**: 480–483.
37. Randolph GJ, Sanchez-Schmitz G, Liebman RM, Schakel K. (2002) The CD16$^+$ (FcgammaRIII$^+$) subset of human monocytes preferentially

becomes migratory dendritic cells in a model tissue setting. *J Exp Med* **196**: 517–527.
38. Randolph GJ, Inaba K, Robbiani DF, *et al.* (1999) Differentiation of phagocytic monocytes into lymph node dendritic cells *in vivo. Immunity* **11**: 753–761.
39. Laouar Y, Welte T, Fu XY, Flavell RA. (2003) STAT3 is required for Flt3L-dependent dendritic cell differentiation. *Immunity* **19**: 903–912.
40. Mohamadzadeh M, Berard F, Essert G, *et al.* (2001) Interleukin 15 skews monocyte differentiation into dendritic cells with features of Langerhans cells. *J Exp Med* **194**: 1013–1020.
41. Chomarat P, Banchereau J, Davoust J, Palucka AK. (2000) IL-6 switches the differentiation of monocytes from dendritic cells to macrophages. *Nat Immunol* **1**: 510–514.
42. Allavena P, Piemonti L, Longoni D, *et al.* (1998) IL-10 prevents the differentiation of monocytes to dendritic cells but promotes their maturation to macrophages. *Eur J Immunol* **28**: 359–369.
43. Delneste Y, Charbonnier P, Herbault N, *et al.* (2003) Interferon-gamma switches monocyte differentiation from dendritic cells to macrophages. *Blood* **101**: 143–150.
44. Rotta G, Edwards EW, Sangaletti S, *et al.* (2003) Lipopolysaccharide or whole bacteria block the conversion of inflammatory monocytes into dendritic cells *in vivo*. *J Exp Med* **198**: 1253–1263.
45. Roodman GD. (1996) Advances in bone biology: the osteoclast. *Endocr Rev* **17**: 308–332.
46. Shalhoub V, Faust J, Boyle WJ, *et al.* (1999) Osteoprotegerin and osteoprotegerin ligand effects on osteoclast formation from human peripheral blood mononuclear cell precursors. *J Cell Biochem* **72**: 251–261.
47. Lacey DL, Timms E, Tan HL, *et al.* (1998) Osteoprotegerin ligand is a cytokine that regulates osteoclast differentiation and activation. *Cell* **93**: 165–176.
48. Fox SW, Fuller K, Bayley KE, *et al.* (2000) TGF-β1 and IFN-gamma direct macrophage activation by TNF-α to osteoclastic or cytocidal phenotype. *J Immunol* **165**: 4957–4963.
49. Zhao Y, Glesne D, Huberman E. (2003) A human peripheral blood monocyte-derived subset acts as pluripotent stem cells. *Proc Nat Acad Sci* **100**: 2426–2431.

50. Kuwana M, Okazaki Y, Kodama H, *et al.* (2003) Human circulating CD14$^+$ monocytes as a source of progenitors that exhibit mesenchymal cell differentiation. *J Leukoc Biol* **74**: 833–845.
51. Hong KM, Burdick MD, Phillips RJ, *et al.* (2005) Characterization of human fibrocytes as circulating adipocyte progenitors and the formation of human adipose tissue in SCID mice. *FASEB J* **19**: 2029–2031.
52. Pilling D, Akbar AN, Girdlestone J, *et al.* (1999) Interferon-β mediates stromal cell rescue of T cells from apoptosis. *Euro Immunolo* **29**: 1041–1050.
53. Buckley CD, Pilling D, Lord JM, *et al.* (2001) Fibroblasts regulate the switch from acute resolving to chronic persistent inflammation. *Trends Immunol* **22**: 199–204.
54. Clark RA. (2001) Fibrin and wound healing. *Ann NY Acad Sci* **936**: 355–367.
55. Majno G. (1998) Chronic inflammation: links with angiogenesis and wound healing. *Am J Pathol* **153**: 1035–1039.
56. Petrakis NL, Davis M, Lucia SP. (1961) The *in vivo* differentiation of human leukocytes into histiocytes, fibroblasts and fat cells in subcutaneous diffusion chambers. *Blood* **17**: 109–118.
57. Labat ML, Bringuier AF, Seebold C, *et al.* (1991) Monocytic origin of fibroblasts: spontaneous transformation of blood monocytes into neo-fibroblastic structures in osteomyelosclerosis and Engelmann's disease. *Biomed Pharmacother* **45**: 289–299.
58. Bringuier AF, Seebold-Choqueux C, Moricard Y, *et al.* (1992) T-lymphocyte control of HLA-DR blood monocyte differentiation into neo-fibroblasts. Further evidence of pluripotential secreting functions of HLA-DR monocytes, involving not only collagen but also uromodulin, amyloid-beta peptide, alpha-fetoprotein and carcinoembryonic antigen. *Biomed Pharmacother* **46**: 91–108.
59. Bucala R, Spiegel LA, Chesney J, *et al.* (1994) Circulating fibrocytes define a new leukocyte subpopulation that mediates tissue repair. *Mol Med* **1**: 71–81.
60. Yang L, Scott PG, Giuffre J, *et al.* (2002) Peripheral blood fibrocytes from burn patients: identification and quantification of fibrocytes in adherent cells cultured from peripheral blood mononuclear cells. *Lab Invest* **82**: 1183–1192.
61. Pilling D, Buckley CD, Salmon M, Gomer RH. (2003) Inhibition of fibrocyte differentiation by serum amyloid P. *J Immunol* **17**: 5537–5546.

62. Chesney J, Metz C, Stavitsky AB, *et al.* (1998) Regulated production of type I collagen and inflammatory cytokines by peripheral blood fibrocytes. *J Immunol* **160**: 419–425.
63. Chesney J, Bacher M, Bender A, Bucala R. (1997) The peripheral blood fibrocyte is a potent antigen-presenting cell capable of priming naive T cells *in situ*. *Proc Natl Acad Sci USA* **94**: 6307–6312.
64. Balmelli C, Ruggli N, McCullough K, Summerfield A. (2005) Fibrocytes are potent stimulators of anti-virus cytotoxic T cells. *J Leukoc Biol* **77**: 923–933.
65. Schmidt M, Sun G, Stacey MA, *et al.* (2003) Identification of circulating fibrocytes as precursors of bronchial myofibroblasts in a asthma. *J Immunol* **171**: 380–389.
66. Mori L, Bellini A, Stacey MA, *et al.* (2005) Fibrocytes contribute to the myofibroblast population in wounded skin and originate from the bone marrow. *Exp Cell Res* **304**: 81–90.
67. Moore BB, Kolodsick JE, Thannickal VJ, *et al.* (2005) CCR2-mediated recruitment of fibrocytes to the alveolar space after fibrotic injury. *Am J Pathol* **166**: 675–684.
68. Grab DJ, Lanners H, Williams WL, Bucala R. (1999) Peripheral blood fibrocytes with foamy virus infection-like morphology. *Hum Pathol* **30**: 1395–1396.
69. Grab D, Salim M, Chesney J. (2002) A role for peripheral blood fibrocytes in Lyme disease? *Med Hypothes* **59**: 1.
70. Cowper SE, Bucala R. (2003) Nephrogenic fibrosing dermopathy: Suspect identified, motive unclear. *Am J Dermatopathol* **25**: 358.
71. Yang L, Scott PG, Dodd C. *et al.* (2005) Identification of fibrocytes in postburn hypertrophic scar. *Wound Rep Regener* **13**: 398–404.
72. Phillips RJ, Burdick MD, Hong K, *et al.* (2004) Circulating fibrocytes traffic to the lungs in response to CXCL12 and mediate fibrosis. *J Clin Invest* **114**: 438–446.
73. Barth PJ, Koster H, Moosdorf R. (2005) $CD34^+$ fibrocytes in normal mitral valves and myxomatous mitral valve degeneration. *Pathol Res Pract* **201**: 301–304.
74. Pilling D, Tucker NM, Gomer RH. (2006) Aggregated IgG inhibits the differentiation of human fibrocytes. *J Leukoc Biol* **79**: 1242–1251.
75. Jain R, Yuen IS, Taphouse CR, Gomer RH. (1992) A density-sensing factor controls development in Dictyostelium. *Genes Dev* **6**: 390–400.

76. Gomer RH. (2001) Not being the wrong size. *Nat Rev Mol Cell Biol* **2**: 48–54.
77. Pilling D, Akbar AN, Shamsadeen N, *et al.* (2000) High cell density provides potent survival signals for resting T-cells. *Cell Mol Biol* **46**: 163–174.
78. Steel DM, Whitehead AS. (1994) The major acute phase reactants: C-reactive protein, serum amyloid P component and serum amyloid A protein. *Immunol Today* **15**: 81–88.
79. Gewurz H, Zhang XH, Lint TF. (1995) Structure and function of the pentraxins. *Curr Opin Immunol* **7**: 54–64.
80. Hutchinson WL, Hohenester E, Pepys MB. (2000) Human serum amyloid P component is a single uncomplexed pentamer in whole serum. *Mol Med* **6**: 482–493.
81. Pepys MB, Booth DR, Hutchinson WL, *et al.* (1997) Amyloid P component. A critical review. *Amyloid* **4**: 274–295.
82. Pepys MB. (1999) Serum amyloid P component (not serum amyloid protein). *Nat Med* **5**: 852–853.
83. Noursadeghi M, Bickerstaff MC, Gallimore JR, *et al.* (2000) Role of serum amyloid P component in bacterial infection: protection of the host or protection of the pathogen. *Proc Natl Acad Sci USA* **97**: 14584–14589.
84. Bickerstaff MC, Botto M, Hutchinson WL, *et al.* (1999) Serum amyloid P component controls chromatin degradation and prevents antinuclear autoimmunity. *Nat Med* **5**: 694–697.
85. Bijl M, Horst G, Bijzet J, *et al.* (2003) Serum amyloid P component binds to late apoptotic cells and mediates their uptake by monocyte-derived macrophages. *Arthritis Rheum* **48**: 248–254.
86. Bharadwaj D, Mold, Markham E, Du Clos TW. (2001) Serum amyloid P component binds to Fc gamma receptors and opsonizes particles for phagocytosis. *J Immunol* **166**: 6735–6741.
87. Mold C, Gresham HD, Du Clos TW. (2001) Serum amyloid P component and C-reactive protein mediate phagocytosis through murine Fc gamma Rs. *J Immunol* **166**: 1200–1205.
88. Marnell LL, Mold C, Volzer MA, *et al.* (1995) C-reactive protein binds to Fc gamma RI in transfected COS cells. *J Immunol* **155**: 2185–2193.
89. Bharadwaj D, Stein MP, Volzer M, *et al.* (1999) The major receptor for C-reactive protein on leukocytes is fcgamma receptor II. *J Exp Med* **190**: 585–590.

90. Du Clos TW, Mold C, Edberg JC, Kimberly RP. (2001) Response: human C-reactive protein does not bind to FcγRIIa on phagocytic cells. *J Clin Invest* **107**: 642.
91. Bodman-Smith KB, Gregory RE, Harrison PT, Raynes JG. (2004) FcgammaRIIa expression with FcgammaRI results in C-reactive protein- and IgG-mediated phagocytosis. *J Leukoc Biol* **75**: 1029–1035.
92. Bodman-Smith KB, Melendez AJ, Campbell I, *et al.* (2002) C-reactive protein-mediated phagocytosis and phospholipase D signalling through the high-affinity receptor for immunoglobulin G (FcgammaRI). *Immunology* **107**: 252–260.
93. Chi M, Tridandapani S, Zhong W, *et al.* (2002) C-reactive protein induces signaling through Fc gamma RIIa on HL-60 granulocytes. *J Immunol* **168**: 1413–1418.
94. Huang ZY, Hunter S, Kim MK, *et al.* (2004) The monocyte Fcgamma receptors FcgammaRI/gamma and FcgammaRIIA differ in their interaction with Syk and with Src-related tyrosine kinases. *J Leukoc Biol* **76**: 491–499.
95. Kim MK, Pan XQ, Huang ZY. (2001) Fc gamma receptors differ in their structural requirements for interaction with the tyrosine kinase Syk in the initial steps of signaling for phagocytosis. *Clin Immunol* **98**: 125–132.
96. Hunter S, Sato N, Kim MK, *et al.* (1999) Structural requirements of Syk kinase for Fcgamma receptor-mediated phagocytosis. *Exp Hematol* **27**: 875–884.
97. Ravetch JV, Kinet JP. (1991) Fc receptors. *Annu Rev Immunol* **9**: 457–492.
98. Ravetch JV, Bolland S. (2001) IgG Fc Receptors. *Annu Rev Immuno* **19**: 275–290.
99. Cochrane CG, Weigle WO, Dixon FJ. (1959) The role of polymorphonuclear leukocytes in the initiation and cessation of the arthus vasculitis. *J Exper Med* **110**: 481–494.
100. Waller M. (1974) IgG hydrolysis in abscesses. I. A study of the IgG in human abscess fluid. *Immunology* **26**: 725–733.
101. Haslett C, Henson PM. (1988) Resolution of inflammation. In: Clark RAF, Henson PM (eds.), *The Molecular and Cellular Biology of Wound Repair*. Plenum Press, New York, pp. 185–211.
102. Shearer JD, Coulter CF, Engeland WC, *et al.* (1997) Insulin is degraded extracellularly in wounds by insulin-degrading enzyme (EC 3.4.24.56). *Am J Physiol Endocrinol Metab* **273**: E657–E664.

103. Greenhalgh DG. (1998) The role of apoptosis in wound healing. *Int J Biochem Cell Biol* **30**: 1019–1030.
104. Dyck RF, Evans DJ, Lockwood CM, *et al.* (1980) Amyloid P-component in human glomerular basement membrane. Abnormal patterns of immunofluorescent staining in glomerular disease. *Lancet* **2**: 606–609.
105. Dyck RF, Lockwood, CM, Kershaw M, *et al.* (1980) Amyloid P-component is a constituent of normal human glomerular basement membrane. *J Exp Med* **152**: 1162–1174.
106. Breathnach SM, Melrose SM, Bhogal B, *et al.* (1981) Amyloid P component is located on elastic fibre microfibrils in normal human tissue. *Nature* **293**: 652–654.
107. Breathnach SM, Melrose SM, Bhogal B, *et al.* (1983) Immunohistochemical studies of amyloid P component distribution in normal human skin. *J Invest Dermatol* **80**: 86–90.
108. Zahedi K. (1996) Characterization of the binding of serum amyloid P to type IV collagen. *J Biol Chem* **271**: 14897–14902.
109. Zahedi K. (1997) Characterization of the binding of serum amyloid P to laminin. *J Biol Chem* **272**: 2143–2148.

Chapter 4

Hematopoietic Origin of Fibrocytes

Amanda C. LaRue* and Makio Ogawa*

Introduction

Fibroblasts in tissues play an important role in growth factor secretion, matrix deposition and matrix degradation and therefore are important in many pathological processes. At the time of tissue injury, fibroblasts are critical to the inflammatory response and its control. Fibroblasts participate in wound healing by producing extracellular matrix proteins, responding to and synthesizing cytokines, chemokines and other mediators of inflammation (for review, see Refs. 1 and 2). In addition, fibroblasts can be activated to become myofibroblasts, which being armed with myosin and alpha smooth muscle actin, exert contractile force to reduce the size of the wound. In contrast to their beneficial role in wound healing, uncontrolled proliferation and/or activation of these cells results in

*Department of Veterans Affairs Medical Center and Division of Experimental Hematology, Department of Medicine, Medical University of South Carolina, Charleston SC 29401. Phone: (843) 789-6712; Fax: (843) 876-5381; Email: laruerc@musc.edu, ogawam@musc.edu.

tissue fibrosis. Fibroblasts and myofibroblasts are also important in the steady state physiology of many organs and tissues. In general, they confer the structural integrity of the tissues and support the proliferation and differentiation of other cell types such as epithelial cells. A number of myofibroblasts with defined tissue-specific functions have been described. For example, contractile myofibroblasts such as glomerular mesangial cells in the kidney, hepatic stellate cells and pericytes function as regulators of blood flow. Readers are referred to reviews (Refs. 3 and 4) for a list of myofibroblasts and their functions.

Fibroblast Precursors

Regarding the origin of tissue fibroblasts and myofibroblasts, it has been long believed that stromal fibroblasts are derived from the recruitment and expansion of resident fibroblasts. However, two types of precursors for fibroblasts with a bone marrow origin, colony forming unit fibroblasts (CFU-F) and fibrocytes, have been identified. More than three decades ago, Friendenstein and colleagues, first demonstrated *in vitro* evidence of fibroblast precursors called CFU-F in the bone marrow.[5,6] In these studies, they cultured non-manipulated bone marrow cells in monolayers and observed discrete colonies of fibroblasts within two weeks of culture.[5] Based on studies in which male and female bone marrow cells were combined in the monolayer culture, the investigators determined that individual fibroblast colonies were the result of clonal outgrowth from a single bone marrow cell. Subsequently, the concept of "epithelial-mesenchymal transition," modeled after embryonic development (reviewed in Ref. 7) and based on the finding that the major population of fibroblasts during renal fibrosis is derived from tubular epithelium,[8] became more widely accepted. Recently, however, a number of *in vivo* transplantation studies have revived the notion of the bone marrow as the source of fibroblasts/myofibroblasts. Using Y-chromosome or green fluorescent protein (GFP) as a marker of donor cells in transplantation studies, investigators have presented evidence that hepatic stellate cells[9]; pericryptal myofibroblasts in the intestine and colon[10]; myofibroblasts in wounded skin[11–13]; tumor formation[13–16]; and fibroblasts in pulmonary fibrosis[17] are derived from bone marrow. In functional studies, transplantation of bone marrow cells reduced the magnitude of liver fibrosis that had been induced with carbon tetrachloride.[18]

Fibrocytes were first described as unique circulating cells that produce matrix proteins and express surface markers associated with the hematopoietic lineage.[19] Fibrocytes have been shown to play a role in numerous pathologies (i.e. wound healing,[11,20] fibrosis[21,22]) and are considered to be fibroblasts/myofibroblasts precursors of bone marrow origin.[11] Fibrocytes comprise 0.1–0.5% of nucleated cells in peripheral blood and express CD34, CD45, CD80, CD86, MHC class II, CD11b and CD13[19–21] and proteins characteristic of fibroblasts, i.e. collagen.[23,24] Fibrocytes in circulation do not express typical surface antigens of monocytes/macrophages. However, populations of cells positive for CD14, an antigen strongly expressed on monocytes and macrophages, can give rise to fibrocytes *in vitro*.[20] While these studies strongly support bone marrow as a source of fibroblasts and fibrocytes, the cellular origin within the bone marrow remained undefined.

Clonal Transplantation

Two types of stem cells, i.e. hematopoietic stem cells (HSCs) and mesenchymal stem cells (MSCs), are thought to be present in the bone marrow. HSCs produce all blood cells and platelets and such tissue cells as osteoclasts and masts cells. MSCs are thought to give rise to various types of mesodermal cells, such as adipocytes, chondrocytes and bone cells (osteocytes and osteoblasts). In order to determine which type of stem cells is the source of fibroblasts/myofibroblasts, we have carried out a series of studies of tissue reconstitution by single HSCs. By transplanting clones derived from single HSCs expressing transgenic enhanced GFP, we found that fibroblasts/myofibroblasts in many organs and tissues are derived from HSCs. This review summarizes these findings and presents perspectives generated by the newly identified differentiation pathway of HSCs.

It was first necessary to generate mice exhibiting high-level, multilineage engraftment from single HSCs in order to study the full differentiation potentials of HSCs. The source of donor cells for these transplantation studies was transgenic GFP mice in C57BL/6 background that were kindly provided by Dr. Okabe, Osaka University.[25] Soon after we began transplantation of single putative HSCs based on the phenotypes of Lin^{-}, Sca-1^{+}, c-kit^{+}, CD34^{-} cells[26] or Lin^{-}, Sca-1^{+}, CD34^{-} SP cells,[27] we found that the success rate of obtaining mice engrafted with single HSCs was too low to be practical. We, therefore, devised a method combining single cell deposition with short-term cell culture.[28,29] Here, Lin^{-}, Sca-1^{+}, c-kit^{+}, CD34^{-} cells or Lin^{-}, Sca-1^{+}, CD34^{-} SP cells from GFP mice were individually cultured for one week in the presence of steel factor and interleukin-11 or a combination of steel factor and granulocyte colony-stimulating factor (G-CSF). Earlier, we had observed that both interleukin 11[30] and G-CSF[31] act on cell cycle-dormant primitive multipotential progenitors and induce cell divisions. Because the majority of HSCs are dormant in cell cycle and do not begin cell division until a few days after initiation of cell culture, transplantation of small clones consisting of 20 or fewer cells after one-week of incubation significantly raised the efficiency of generating mice that were engrafted with single HSCs. Another important issue was selection of the type of radio-protective cells given to the lethally irradiated recipients. Radio-protective cells of the recipients' type are necessary for protection of the recipients from the complications of radiation-induced bone marrow aplasia, which lasts for about two weeks after radiation. Ideally, the radio-protective cells consist of hematopoietic progenitors that produce mature granulocytes and platelets for about two weeks but are devoid of HSCs that will compete with the GFP^{+} clones and reduce the engraftment levels. Although "compromised" bone marrow cells that have undergone two previous cycles of transplantation[32] are often used for this purpose, we chose 500 Lin^{-}, Sca-1^{+}, c-kit^{+}, CD34^{+}cells,[26] because this population contains almost no HSCs. Cultured clones of GFP^{+} cells, together with 500 Lin^{-}, Sca-1^{+}, c-kit^{+}, CD34^{+} bone marrow cells from normal non-GFP recipient mice, were individually transplanted to lethally irradiated non-GFP recipients. Two months to one year after cell transplantation, nucleated blood cells from these mice were analyzed for hematopoietic engraftment and only the mice revealing high-level multilineage engraftment by donor GFP^+ cells were selected for analysis of tissue reconstitution. We also carried out transplantation of 100 non-cultured Lin^-, Sca-1^+, c-kit^+, $CD34^-$ cells and made similar observations to those seen in clonally engrafted mice. These findings excluded the possibility that the observed results from clonally engrafted mice are artifacts of short-term cell cultures. In most of the studies described below, we excluded the possibility of cell fusions by carrying out male-to-male or female-to-male transplantation and analyzing the number of Y-chromosomes in the GFP^+ cells.

HSC Origin of Fibroblasts

The first type of HSC-derived myofibroblasts we detected was glomerular mesangial cells of the kidney.[28] The location in the kidney and the morphology of the GFP^+ cells suggested that the cells were mesangial cells. This identification was confirmed by the ability of the GFP^+ cells to contract upon exposure to angiotensin II. Next, we discovered that brain microglial cells and perivascular cells are of HSC origin and demonstrated that induction of stroke by ligation of the middle cerebral artery strongly enhanced the recruitment of the GFP^+ microglial cells to the injury site.[29] The morphological and immunohistochemical properties of the GFP^+ perivascular cells were consistent with the cells being pericytes rather than endothelial cells. We then identified HSC-derived GFP^+ fibroblasts associated with transplantable murine melanoma or Lewis lung carcinoma.[33] The GFP^+ cells were demonstrated to be fibroblasts by their distinct morphology and expression of procollagen mRNA. In addition, a subpopulation of GFP^+ fibroblastic cells were positive for smooth muscle actin, indicating they were myofibroblasts. Also prevalent in the specimens were GFP^+ pericyte-like perivascular cells admixed with tumor cells. Similar to our findings in the brain, simultaneous staining for CD31 expression clearly established that the perivascular cells were not endothelial cells. Recently, we found that fibrocytes and other unidentified mesenchymal-type cells in the spiral ligament of the inner ear are of HSC origin.[34] Inner ear fibrocytes are known to play a critical role in the homeostasis of inner ear ion and fluid channels and are important for the health of the inner ear hair cells. These fibrocytes are classified into five types based on location, morphology and histochemical properties. GFP^+ cells were seen among all five types of fibrocytes.[34] Most recently, we found fibroblasts/myofibroblasts of HSC origin in the adult heart valves.[35] Finally, an abstract was presented at the annual meeting of the American Society of Hematology, demonstrating myofibroblasts of HSC origin that are recruited to the injury site following induction of myocardial infarction.[36]

In addition to these *in vivo* studies demonstrating an HSC origin for tissue fibroblasts/myofibroblasts, we have also succeeded in culturing fibroblasts from GFP^+ bone marrow cells of clonally engrafted mice.[37] GFP^+ bone marrow cells from these mice, incubated in fibronectin-coated tissue culture dishes or flasks in the presence of 10% mouse serum and 10% fetal bovine serum, generated adherent cells exhibiting the morphology of fibroblasts described three decades ago by Friedenstein and associates.[5,6] The GFP^+cells showed spindle-shaped or pleomorphic cytoplasm and prominent clear nuclei. They also expressed mRNAs for procollagen 1αI, fibronectin, vimentin and discoidin domain receptor type 2 (DDR2). Time-course flow cytometric analyses of the cultured GFP^+ bone marrow cells revealed gradual expression of collagen I and DDR2 and concomitant loss of CD45 during three weeks of incubation. As to the precursors of fibroblasts, both CFU-F[5,38] and peripheral blood fibrocytes were derived from the bone marrow of the clonally engrafted mice.[37] With regard to CFU-F, we demonstrated that a culture of mononuclear cells prepared from the bone marrow of clonally engrafted mice resulted in colonies consisting of more than 50 GFP^+ adherent cells in all the culture plates. The majority of these cells were fibroblast-like and exhibited spindle-shaped or polygonal cytoplasm and clear ovoid nuclei, a morphology consistent with that originally described by Friedenstein.[5,6] We also characterized circulating fibrocytes in clonally engrafted mice by culturing nucleated blood cells in fibronectin-coated dishes for 10 to 14 days. The appearance of GFP^+spindle-shaped or polygonal cells was detected by the 7th day of cultivation. By day 14 of incubation, most of the GFP^+ cells expressed collagen-I and DDR2 as assessed by flow cytometry. Only one-third of the GFP^+ cells expressed CD45. This cell culture study was consistent with the results of *in vivo* transplantation studies described above and strongly suggested that most, if not all, fibrocytes and fibroblasts/myofibroblasts are derived from HSCs.

Perspectives

There are earlier studies suggesting a close relationship between fibroblasts and the hematopoietic system, in particular the macrophage lineage. A population of cells in CFU-F-derived mouse colonies was positive for Mac-1 and F4/80.[39] When $CD34^+$ human peripheral blood or cord blood cells were transplanted into non-obese diabetic/severe combined immune deficiency (NOD/SCID) mice and showed human cell engraftment, the bone marrow was shown to contain $5B5^+$ human fibroblasts and express human proline hydroxylase-alpha mRNA, an enzyme required for collagen synthesis by fibroblasts.[40] Human peripheral blood cells that were positive for CD14, a surface protein preferentially expressed on monocytes and macrophages, gave rise to fibrocytes when co-cultured with T-cells.[20] When we examined possible correlations between HSC-derived GFP^+glomerular mesangial cells and hematopoietic lineages in clonally engrafted mice, GFP^+mesangial cells were detected only in the mice expressing Mac-1/$Gr\text{-}1^+$cells in the blood.[41] In addition, humoral regulation of stromal cells suggests their relation to macrophage lineage. Two groups of investigators observed that M-CSF supports proliferation of fibroblastic stromal cells in culture.[42–45] Although most of these studies are only correlative in nature, they nonetheless suggest closeness between the macrophage lineage and fibroblasts and are consistent with our series of observations *in vivo* based on clonal HSC transplantation.

The concept that fibroblasts/myofibroblasts are derived from HSCs may lead to new avenues of therapy for injury-related disorders. A number of investigators have noted that administration of G-CSF reduces the size of stroke and improves functional outcome in studies of murine models.[46–48] We reported that contractile brain microglial cellsmicroglial cells are derived from HSCs and that they are dramatically increased in number following induction of stroke.[29] G-CSF is known to mobilize HSCs to the circulation. Therefore, recruitment by G-CSF of microglial cell precursors to the site of injury and their subsequent differentiation must be the reason for the reduction in the size of stroke. Some of the controversies surrounding the therapeutic role of HSCs in myocardial infarction may be clarified by our observations of HSC-derived fibroblasts/myofibroblasts. Previously, using a murine model of heart attack, Orlic and his associates reported that transplanted bone marrow cells regenerate infarcted myocardium[49] and that mobilization of the primitive bone marrow cells by G-CSF to the site of infarct confers therapeutic effects.[50] Subsequently, there has been much discussion as to the nature of the HSC-derived cells engrafting the injury site.[51–53] The current consensus appears to be that very few cardiomyocytes show donor markers and that they are the products of cell fusions between donor cells and recipient cardiac cells.[54] The majority of cells of donor origin at the injury site appear to be fibroblasts/myofibroblasts.[36] Similar to the findings in the murine stroke model, the apparent therapeutic effects of G-CSF may be mobilization of fibroblasts/myofibroblasts to the site of myocardial infarction and consequent reduction in the size of the scar.[36] These observations in the mouse models of stroke and myocardial infarction may suggest new therapeutic approaches to injuries of many organs and tissues.

Finally, the discovery of an HSC origin of fibroblasts/myofibroblasts raises serious questions regarding the current model of stem cell systems in the bone marrow. It has been generally believed that there are two types of stem cells in the bone marrow, i.e. HSCs and MSCs. HSCs produce blood cells and some cells in the tissues such as mast cells and osteoclasts. MSCs give rise to a number of mesenchymal cells, including adipocytes, chondrocytes and osteocytes and CFU-F are thought to be their precursors.[55–60] While the repertoire of HSC and MSC potential had been thought to be distinct and separate from each other, several studies have begun to question this distinction. A single SP cell and 3000 SP cells generated osteoblasts in culture and *in vivo*, respectively.[61] Our observation that CFU-F, which had been thought to be progenitors of mesenchymal cells, are also derived from HSCs[37] further blurring the line between HSCs and MSCs. MSCs are far less clearly defined than HSCs, as pointed out by a recent in depth review.[62] Indeed, despite significant academic and commercial interest and current ongoing clinical trials, the exact phenotype of MSCs is not known and most of the studies have been performed *in vitro* on "fibroblastic" cells. While some transplantation studies have been performed with MSCs, identification of the reconstituted cells was not thoroughly presented. The concept that fibroblasts/myofibroblasts are of HSC origin may further indicate that other types of mesenchymal cells may also have their origin in HSCs.

Acknowledgments

This work was supported by the office of Research and Development, Medical Research Services, Department of Veterans Affairs (MO and ACL) and by an NIH grant, RO1 HL69123 (MO).

References

1. Eckes B, Zigrino P, Kessler D, *et al.* (2000) Fibroblast-matrix interactions in wound healing and fibrosis. *Matrix Biol* **19**: 325–332.
2. Gabbiani G. (2003) The myofibroblast in wound healing and fibrocontractive diseases. *J Pathol* **200**: 500–503.
3. Powell DW, Mifflin RC, Valentich JD, *et al.* (1999) Myofibroblasts. I. Paracrine cells important in health and disease. *Am J Physiol* **277**: C1–C9.
4. Tomasek JJ, Gabbiani G, Hinz B, *et al.* (2002) Myofibroblasts and mechano-regulation of connective tissue remodelling. *Nat Rev Mol Cell Biol* **3**: 349–363.

5. Friedenstein AJ, Chailakhjan RK, Lalykina KS. (1970) The development of fibroblast colonies in monolayer cultures of guinea-pig bone marrow and spleen cells. *Cell Tissue Kinet* **3**: 393–403.
6. Luria EA, Panasyuk AF, Friedenstein AY. (1971) Fibroblast colony formation from monolayer cultures of blood cells. *Transfusion* **11**: 345–349.
7. Hay ED. (2005) The mesenchymal cell, its role in the embryo, and the remarkable signaling mechanisms that create it. *Dev Dyn* **233**: 706–720.
8. Iwano M, Plieth D, Danoff TM, *et al.* (2002) Evidence that fibroblasts derive from epithelium during tissue fibrosis. *J Clin Invest* **110**: 341–350.
9. Forbes SJ, Russo FP, Rey V, *et al.* (2004) A significant proportion of myofibroblasts are of bone marrow origin in human liver fibrosis. *Gastroenterology* **126**: 955–963.
10. Brittan M, Hunt T, Jeffery R, *et al.* (2002) Bone marrow derivation of pericryptal myofibroblasts in the mouse and human small intestine and colon. *Gut* **50**: 752–757.
11. Mori L, Bellini A, Stacey MA, *et al.* (2005) Fibrocytes contribute to the myofibroblast population in wounded skin and originate from the bone marrow. *Exp Cell Res* **304**: 81–90.
12. Direkze NC, Forbes SJ, Brittan M, *et al.* (2003) Multiple organ engraftment by bone-marrow-derived myofibroblasts and fibroblasts in bone-marrow-transplanted mice. *Stem Cells* **21**: 514–520.
13. Ishii G, Sangai T, Sugiyama K, *et al.* (2005) *In vivo* characterization of bone marrow-derived fibroblasts recruited into fibrotic lesions. *Stem Cells* **23**: 699–706.
14. Ziegelhoeffer T, Fernandez B, Kostin S, *et al.* (2004) Bone marrow-derived cells do not incorporate into the adult growing vasculature. *Circ Res* **94**: 230–238.
15. Rajantie I, Ilmonen M, Alminaite A, *et al.* (2004) Adult bone marrow-derived cells recruited during angiogenesis comprise precursors for periendothelial vascular mural cells. *Blood* **104**: 2084–2086.
16. Direkze NC, Hodivala-Dilke K, Jeffery R, *et al.* (2004) Bone marrow contribution to tumor-associated myofibroblasts and fibroblasts. *Cancer Res* **64**: 8492–8495.
17. Hashimoto N, Jin H, Liu T, *et al.* (2004) Bone marrow-derived progenitor cells in pulmonary fibrosis. *J Clin Invest* **113**: 243–252.
18. Sakaida I, Terai S, Yamamoto N, *et al.* (2004) Transplantation of bone marrow cells reduces CCl4-induced liver fibrosis in mice. *Hepatology* **40**: 1304–1311.

19. Bucala R, Spiegel LA, Chesney J, *et al.* (1994) Circulating fibrocytes define a new leukocyte subpopulation that mediates tissue repair. *Mol Med* **1**: 71–81.
20. Abe R, Donnelly SC, Peng T, *et al.* (2001) Peripheral blood fibrocytes: Differentiation pathway and migration to wound sites. *J Immunol* **166**: 7556–7562.
21. Schmidt M, Sun G, Stacey MA, *et al.* (2003) Identification of circulating fibrocytes as precursors of bronchial myofibroblasts in asthma. *J Immunol* **171**: 380–389.
22. Phillips RJ, Burdick MD, Hong K, *et al.* (2004) Circulating fibrocytes traffic to the lungs in response to CXCL12 and mediate fibrosis. *J Clin Invest* **114**: 438–446.
23. Chesney J, Bacher M, Bender A, *et al.* (1997) The peripheral blood fibrocyte is a potent antigen-presenting cell capable of priming naive T cells *in situ*. *Proc Natl Acad Sci USA* **94**: 6307–6312.
24. Chesney J, Metz C, Stavitsky AB, *et al.* (1998) Regulated production of type I collagen and inflammatory cytokines by peripheral blood fibrocytes. *J Immunol* **160**: 419–425.
25. Okabe M, Ikawa M, Kominami K, *et al.* (1997) 'Green mice' as a source of ubiquitous green cells. *FEBS Lett* **407**: 313–319.
26. Osawa M, Hanada K, Hamada H, *et al.* (1996) Long-term lymphohematopoietic reconstitution by a single CD34-low/negative hematopoietic stem cell. *Science* **273**: 242–245.
27. Matsuzaki Y, Kinjo K, Mulligan RC, *et al.* (2004) Unexpectedly efficient homing capacity of purified murine hematopoietic stem cells. *Immunity* **20**: 87–93.
28. Masuya M, Drake CJ, Fleming PA, *et al.* (2003) Hematopoietic origin of glomerular mesangial cells. *Blood* **101**: 2215–2218.
29. Hess DC, Abe T, Hill WD, *et al.* (2004) Hematopoietic origin of microglial and perivascular cells in brain. *Exp Neurol* **186**: 134–144.
30. Musashi M, Clark SC, Sudo T, *et al.* (1991) Synergistic interactions between interleukin-11 and interleukin-4 in support of proliferation of primitive hematopoietic progenitors of mice. *Blood* **78**: 1448–1451.
31. Ikebuchi K, Clark SC, Ihle JN, *et al.* (1988) Granulocyte colony-stimulating factor enhances interleukin 3-dependent proliferation of multipotential hemopoietic progenitors. *Proc Natl Acad Sci USA* **85**: 3445–3449.

32. Harrison DE, Astle CM, Delaittre JA. (1978) Loss of proliferative capacity in immunohemopoietic stem cells caused by serial transplantation rather than aging. *J Exp Med* **147**: 1526–1531.
33. LaRue AC, Masuya M, Ebihara Y, *et al.* (2006) Hematopoietic origins of fibroblasts: I. *In vivo* studies of fibroblasts associated with solid tumors. *Exp Hematol* **34**: 208–218.
34. Lang H, Ebihara Y, Schmiedt RA, *et al.* (2006) Contribution of bone marrow hematopoietic stem cells to adult mouse inner ear: Mesenchymal cells and fibrocytes. *J Comp Neurol* **496**: 187–201.
35. Visconti RP, Ebihara Y, LaRue AC, *et al.* (2006) An *in vivo* analysis of hematopoietic stem cell potential: Hematopoietic origin of cardiac valve interstitial cells. *Circ Res* **98**: 690–696.
36. Kawada H, Fujita J, Tsuma M, *et al.* (2005) Cardiac myofibroblasts of hematopoietic origin are mobilized by G-CSF and contribute to cardiac repair after myocardial infarction. American Society of Hematology, Atlanta, GA. *Blood* **106**: 484a–485b.
37. Ebihara Y, Masuya M, Larue AC, *et al.* (2006) Hematopoietic origins of fibroblasts: II. *In vitro* studies of fibroblasts, CFU-F, and fibrocytes. *Exp Hematol* **34**: 219–229.
38. Friedenstein AJ, Gorskaja JF Kulagina NN. (1976) Fibroblast precursors in normal and irradiated mouse hematopoietic organs. *Exp Hematol* **4**: 267–274.
39. Penn PE, Jiang DZ, Fei RG, *et al.* (1993) Dissecting the hematopoietic microenvironment. IX. Further characterization of murine bone marrow stromal cells. *Blood* **81**: 1205–1213.
40. Goan SR, Junghahn I, Wisslerm M, *et al.* (2000) Donor stromal cells from human blood engraft in NOD/SCID mice. *Blood* **96**: 3971–3978.
41. Abe T, Fleming PA, Masuya M, *et al.* (2005) Granulocyte/macrophage origin of glomerular mesangial cells. *Int J Hematol* **82**: 115–118.
42. Deryugina EI, Ratnikov BI, Bourdon MA, *et al.* (1994) Clonal analysis of primary marrow stroma: Functional homogeneity in support of lymphoid and myeloid cell lines and identification of positive and negative regulators. *Exp Hematol* **22**: 910–918.
43. Deryugina EI, Ratnikov BI, Bourdon MA, *et al.* (1995) Identification of a growth factor for primary murine stroma as macrophage colony-stimulating factor. *Blood* **86**: 2568–2578.
44. Yamada M, Suzu S, Akaiwa E, *et al.* (1997) Properties of primary murine stroma induced by macrophage colony-stimulating factor. *J Cell Physiol* **173**: 1–9.

45. Yamada M, Suzu S, Tanaka-Douzono M, *et al.* (2000) Effect of cytokines on the proliferaton/differentiation of stroma-initiating cells. *J Cell Physiol* **184**: 351–355.
46. Six I, Gasan G, Mura E, Bordet R. (2003) Beneficial effect of pharmacological mobilization of bone marrow in experimental cerebral ischemia. *Eur J Pharmacol* **458**: 327–328.
47. Shyu WC, Linsz, Yang HI, *et al.* (2004) Functional recovery of stroke rats induced by granulocyte colony-stimulating factor-stimulated stem cells. *Circulation* **110**: 1847–1854.
48. Gibson CL, Bath PM, Murphy SP. (2005) G-CSF reduces infarct volume and improves functional outcome after transient focal cerebral ischemia in mice. *J Cereb Blood Flow Metab* **25**: 431–439.
49. Orlic D, Kajstura J, Chimenti S, *et al.* (2001) Bone marrow cells regenerate infarcted myocardium. *Nature* **410**: 701–705.
50. Orlic D, Kajstura J, Chimenti S, *et al.* (2001) Mobilized bone marrow cells repair the infarcted heart, improving function and survival. *Proc Natl Acad Sci USA* **98**: 10344–10349.
51. Orlic D, Kajstura J, Chimenti S, *et al.* (2003) Bone marrow stem cells regenerate infarcted myocardium. *Pediatr Transplant* **7**(3): 86–88.
52. Balsam LB, Wagers AJ, Christensen JL, *et al.* (2004) Haematopoietic stem cells adopt mature haematopoietic fates in ischaemic myocardium. *Nature* **428**: 668–673.
53. Murry CE, Soonpaa MH, Reinecke H, *et al.* (2004) Haematopoietic stem cells do not transdifferentiate into cardiac myocytes in myocardial infarcts. *Nature* **428**: 664–668.
54. Nygren JM, Jovinge S, Breitbach M, *et al.* (2004) Bone marrow-derived hematopoietic cells generate cardiomyocytes at a low frequency through cell fusion, but not transdifferentiation. *Nat Med* **10**: 494–501.
55. Ashton BA, Allen TD, Howlett CR, *et al.* (1980) Formation of bone and cartilage by marrow stromal cells in diffusion chambers *in vivo*. *Clin Orthop Relat Res*: 294–307.
56. Friedenstein AJ, Chailakhyan RK, Gerasimov UV. (1987) Bone marrow osteogenic stem cells: *in vitro* cultivation and transplantation in diffusion chambers. *Cell Tissue Kinet* **20**: 263–272.
57. Prockop DJ. (1997) Marrow stromal cells as stem cells for non-hematopoietic tissues. *Science* **276**: 71–74.

58. Pittenger MF, Mackay AM, Beck SC, *et al.* (1999) Multilineage potential of adult human mesenchymal stem cells. *Science* **284**: 143–147.
59. Verfaillie CM, Schwartz R, Reyes M, *et al.* (2003) Unexpected potential of adult stem cells. *Ann N Y Acad Sci* **996**: 231–234.
60. Gregory CA, Prockop DJ, Spees JL. (2005) Non-hematopoietic bone marrow stem cells: molecular control of expansion and differentiation. *Exp Cell Res* **306**: 330–335.
61. Olmsted-Davis EA, Gugala Z, Camargo F, *et al.* (2003) Primitive adult hematopoietic stem cells can function as osteoblast precursors. *Proc Natl Acad Sci USA* **100**: 15877–15882.
62. Javazon EH, Beggs KJ, Flake AW. (2004) Mesenchymal stem cells: Paradoxes of passaging. *Exp Hematol* **32**: 414–425.

Chapter 5

The Role of Fibrocytes in Post-burn Hypertrophic Scarring

JianFei Wang, Yaujiong Wu, Abelardo Medina, Paul. G. Scott and Edward E. Tredget*

Introduction

Unlike fetal wound healing which is characterized by regenerating normal epidermis and dermis with restoration of the extracellular matrix architecture, strength and function without scarring,[1] adult wound healing results in scar formation and epidermal appendages do not regenerate. The scar remains a connective tissue product where the collagen matrix has been poorly reconstituted, in dense parallel bundles, unlike the mechanically efficient basket-wave meshwork of collagen in normal dermis.[2] Scar formation and contraction lead to defective growth, impaired function, and an unpleasant cosmetic appearance.[3] Post-burn hypertrophic scarring represents one form of aberrant wound healing, characterized by a raised, erythematous, pruritic and inelastic mass of tissue.[4] If left untreated it may undergo a reorganization of the collagen within its

*Corresponding author: 2D3.81 WMC, 8440-112 Street, University of Alberta, Edmonton, Alberta, Canada T6G 2B7. Tel.: 780-407-6979; Fax: 780-407-7394. E-mail: etredget@ualberta.ca

dermal matrix, leading to the development of contractures and thus adding functional impairment to the discomfort and cosmetic problems already suffered by the recovering patients (Fig. 1).[5] Although it is known that wound healing involves a complex interplay of cellular, humoral and biochemical factors, the mechanism of hypertrophic scar formation is still not fully understood. Here we review recent research on the contribution of fibrocytes to the altered structure and composition of extracellular matrix, the elevated levels of profibrogenic factors TGF-β and CTGF, dysregulated apoptosis, polarized immune response, increased myofibroblast population, and their potential interaction with dermal fibroblasts.

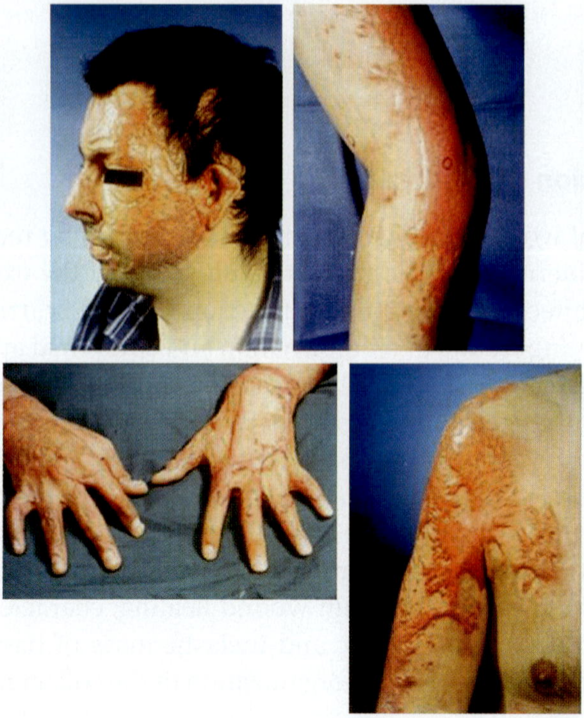

Fig. 1. Hypertrophic scarring in a 34-year-old white man, 8 months following a 60% total body surface area burn involving the face, upper extremeties and hands. (From Scott, PG *et al.*, 2000, with permission.)

Altered Structure and Composition of the Extracellular Matrix of Hypertrophic Scars

In a fibrous connective tissue such as the dermis, collagen is the major structural protein and provides mechanical strength to the skin. Type I collagen represents approximately 80% of the collagen present, with the remainder being primarily type III (~10%). Following wounding, there is a shift in the relative proportion of these two types of collagen, with an increased amount of type III which then decreases with remodeling.[6] Collagen is an exceedingly tough fibrous protein that is secreted as procollagen by the fibroblast into the extracellular environment, where it undergoes processing by enzymes that cleave the propeptide portions of the polypeptides to form mature collagen.[6] The triple-helical collagen macromolecules assemble into fibrils that are then stabilized by covalent crosslinking. In normal dermis, these fibrils further assemble into fibers and fiber-bundles. While the tensile strength of the skin is provided by collagen, the ability to recoil after transient stretching is provided by elastic fibers.[6] Elastin is co-deposited with collagen by fibroblasts into the dermis. Other major proteins found in the dermis are the proteoglycans, which have polysaccharide glycosaminoglycan chains attached to a protein core. Glycosaminoglyans are hydrophilic and capable of absorbing up to 1000 times their volume of water to form an aqueous gel.[6] Major glycosaminoglycans found in the dermis are hyaluronic acid, chondroitin sulfate, dermatan sulfate and heparan sulfate. This hydrated material probably helps to maintain the appropriate water balance for metabolic needs and may promote cell migration, attachment and differentiation. Proteoglycans are more abundant in the papillary dermis. Fibronectin is the major adhesive glycoprotein of the dermis. During wound healing it functions as a provisional matrix to facilitate the migration, adhesion, spreading and chemotaxis of cells and to promote re-epithelialization.[6]

Interaction with extracellular matrix (ECM) can control both cell phenotype and behavior.[3,4] Interaction between the individual components of the ECM and specific cell surface molecules can initiate a cascade of signal transduction events leading to varied cellular responses. The ECM is also important as a reservoir for growth factors and cytokines, and components of the ECM can interact with cytokines in either a synergistic or an antagonistic fashion.

In hypertrophic scars resulting from thermal injury, the ECM constituents differ both qualitatively and quantitatively from those of the skin or normotrophic or mature scar.[5] Increased collagen content is usually considered the hallmark of hypertrophic scar; however, the proportion of collagen on a dry-weight basis is actually 30% lower than in normal dermis or mature scar.[7] This difference is explained by the increased content of proteoglycan and glycoproteins (see below). Collagen in post-burn hypertrophic scar is primarily of the same genetic type (I) as that in normal dermis but with higher proportions of type III (~33%). The undesirable physical properties of hypertrophic scar are a direct consequence of the deranged organization of the dermis in which the collagen fibrils are narrower, more widely spaced and often arranged in whorls or nodules, rather than as the coarse fibers and fiber bundles running parallel to the surface, as seen in normal skin or mature scar.[8] The temporary absence of elastin[8] has been suggested to contribute to the clinical finding of hardness and inelasticity of hypertrophic scars.

Disorganization of the collagen may result from abnormalities in type and/or amounts of proteoglycans and glycoproteins. Hypertrophic scar is known to contain more water than normal dermis or mature scar, higher concentrations of the glycosaminoglycan components uronic acid and hexosamine and more fibronectin.[6] The glycosaminoglycans differ qualitatively from those in normal dermis: chondroitin sulphate which is normally barely detectable is readily demonstrated in hypertrophic scar[6] and dermatan sulphate, the major glycosaminoglycan of the normal dermis, is virtually absent.[9] Our own immunohistochemical[10] and chemical[7] studies showed the presence of abnormally high levels of biglycan and versican in hypertrophic scar and the virtual absence of decorin, while the levels of these proteoglycans were normal in mature post-burn scars. Isolated fibroblasts from post-burn hypertrophic scar tissue synthesize less decorin than normal dermal fibroblasts.[11] Aberrant proteoglycan metabolism may therefore be a significant factor contributing to the altered physical properties of hypertrophic scars and the maturation of post-burn scars may be dependent on a return of the relative proportions and concentrations of proteoglycans to those that are characteristic of normal dermis.

A Th2 Polarized Immune Response in Hypertrophic Scar

Immunologic factors have been suggested to play a central role in the disruption of the normal process of wound healing and tissue remodeling.[12] It has long been recognized that hypertrophic scars are greatly infiltrated with lymphocytes, often forming perivascular cuffs. Examination of skin-infiltrating cells by means of immunohistochemistry showed the presence of more immunocompetent CD3+ T cells in active hypertrophic scars than in normotrophic scars.[13,14] CD4+ T cells predominated in the dermis as well as in the epidermis in hypertrophic scar,[15] whereas CD8+ T cells were less represented.[13] About 70% of T-lymphocytes present in active hypertrophic scar were activated (expressing HLA-DR and IL-2R). This is significantly higher than the levels present in remission-phase hypertrophic scars and normotrophic or mature scars.[13] In addition, there were increased numbers of macrophages and Langerhan's cells, but no B cells were seen in hypertrophic scar tissue.[13] Central to the immune hypothesis of hypertrophic scars is that some of the T-cell lymphokines act on keratinocytes, fibroblasts and other cell types to induce changes characteristic of these scars. The presence and close proximity of activated T lymphocytes and antigen-presenting cells of various phenotypes in both the epidermis and dermis of hypertrophic tissues provides strong circumstantial evidence of an important local immune response. However, the manner in which T cells achieve and maintain their activated state in hypertrophic tissues is not yet known, and both antigen-dependent and independent mechanisms may contribute.

It has long been suggested that T cells (CD4+) are major immunoregulators in wound healing, functioning by the production of cytokines after activation by macrophages, Langerhans cells, or other antigen presentation cells (APCs) in the context of a major histocompatibility complex molecule, which binds to the T cell receptor. Mosmann *et al.*[16] defined in mice two subsets of CD4+ T cells that produce distinct groups of cytokines. Th1 clones express IL-2, IFN-γ and lymophotoxin, and promote IgG2a antibody production by B cells, and are principally mediators of cell mediated immunity; whereas, Th2 clones express IL-4, IL-5 and IL-10, and are associated with antibody mediated immunity, IgE and IgG1 production, eosinophilia and mast cells.[17] Both types of Th cell, however, are thought to arise from a common precursor. The presence of IL-4 in the early stages of stimulation leads to strong polarization of T cells toward the Th2 phenotype and secretion of high levels of IL-4 upon re-stimulation.[18] IL-4, together with IL-10, inhibits the synthesis of IFN-γ and other Th1 cytokines.[19] On the other hand, IL-12 is very effective in inducing IFN-γ, but is inhibited by the Th2 cytokines, IL-4 and IL-10.[19,20] *In vivo* evidence exists to suggest that the murine patterns of cytokine production occur in humans and are involved in the pathogenesis of a number of disorders such as systemic lupus erythematosus,[21] systemic scleroderma,[22] and atopic dermatitis.[23]

Reduced Th1 cytokines, IL-12 and IFN-γ production and increased Th2 cytokines, have been reported during the open wound phase of injury in animal models and in humans with acute burn injury.[24,25] In our longitudinal studies in human patients with hypertrophic scar, within one month of injury, few IFN-γ positive T cells were found in association with low IL-12 and absent IFN-γ cytokine levels in the serum; whereas IL-4 positive Th2 cells were significantly increased as compared to normal controls by two months post-injury. In burn patients with hypertrophic scar, IL-10 serum levels were also significantly increased early after burn injury as compared to normal volunteers and to a subset of burn patients who did not develop hypertrophic scar, before returning to normal levels after 6 months. mRNA for IFN-γ was detected only in activated peripheral blood mononuclear cells in normal or patients without hypertrophic scar, but was undetectable in hypertrophic scar patients; whereas IL-4 mRNA levels were increased in the PBMC of burn patients with hypertrophic scar. In hypertrophic scar tissue IL-4 mRNA was increased; whereas, IFN-γ mRNA was reduced as compared to normal skin and mature scar. Increased CD3+ and CD4+ cells were present in hypertrophic scar as compared to normal skin and were co-expressed with the fibrogenic cytokine TGF-β. These longitudinal studies in human patients suggest that scar formation in the skin is associated with a polarized Th2 systemic response to injury that leads to increased T cells and their Th2 fibrogenic cytokines in tissue, and the subsequent development of fibrotic hypertrophic scar.[15]

Dysregulated Apoptosis in Hypertrophic Scar

Wound healing involves a localized, coordinated increase in neutrophils, macrophages, lymphocytes, endothelial cells and fibroblasts. This prepares the wound for repair and the deposition of new matrix leading to maturation of the scar. These processes are sequential, and when one type of cell completes its mission, it must be eliminated from the wound before the next phase of healing. Tissue repair involves a transition from a highly cellular and vascular wound to a relatively acellular and avascular scar. The total number of cells gradually decreases in a mature scar. Apoptosis has been reported to play a key role in cell elimination during wound healing.[26] Apoptosis appears to have two distinct roles in the healing wound. The first involves the removal of inflammatory cells, signaling the end of the inflammatory phase and the start of the proliferative phase. The second role involves down-regulation of fibroblasts and collagen deposition, as required for wound maturation.[27] As inflammation is essential for healing, cellular infiltration must be sufficient for normal, early progression of repair. Later, the inflammatory process must be turned down or off, as the wound is re-epithelialized. Failure to end inflammation and matrix deposition will result in abnormal wound healing.

Apoptosis in the Resolution of Inflammation

Tissue injury is intimately associated with the onset of acute inflammation and the arrival of polymorphonuclear leukocytes (PMNs) attracted to the wound site by chemotactic signals such as growth factors released by degranulating platelets, bacterial proteins with formylated N-terminal methionine and peptide byproducts of fibrin and matrix proteolysis.[28] Increased PMN infiltration is seen in early wound healing. If there is no substantial wound contamination, PMNs will be eliminated within a few days. PMNs isolated from the blood are normally short-lived (8–20 hrs) and have been shown to undergo spontaneous apoptosis in culture.[29] However, this short lifespan can increase several-fold once the PMNs enter infected or inflamed tissue.[30] Under normal conditions, monocytes/macrophages also invade the wound. After an initial influx immediately following injury, the PMN number then steadily declines as a result of spontaneous apoptosis. Apoptotic PMNs are recognized and ingested by macrophages. Apoptosis is observed in PMN disappearance, and plays an essential role in resolving acute inflammation in a pig skin wound healing model.[31] Macrophages retrieved from rat wounds with a polyvinyl alcohol sponge, where the disappearance of PMN coincides with the establishment of a macrophage-dominant infiltrate, were able to ingest apoptotic, but not viable, wound PMN.[32] These studies suggest that apoptosis of neutrophils and their subsequent phagocytosis by macrophages, play a role in the resolution of inflammation. Pathological scar formation resulting from thermal injury is closely associated with protracted inflammation.[3,4] Bacterial contamination creates an environment that prolongs PMN lifespan and it is well known that prolonged inflammation leads to hypertrophic scarring.[33]

Delayed Fibroblast and Myofibroblast Apoptosis in Hypertrophic Scar

One of the features of hypertrophic scar is contracture and myofibroblasts are transiently found at the site of tissue injury.[34] By secreting extracellular matrix proteins and by promoting the contraction of the granulation tissue through the expression of alpha-smooth muscle actin (αSMA), these cells are believed by some investigators to be essential for wound closure. Increased numbers of myofibroblasts were found in hypertrophic scar.[35,36] Delayed apoptosis of myofibroblasts has long been suggested in hypertrophic scar. The bcl-2 proto-oncogene, whose protein product protects cells from apoptosis, was strongly up-regulated in hypertrophic scar[37]; however, the death receptor ligand Fas was down-regulated, both in hypertrophic scar and in the fibroblasts cultured from it.[38] The transcription factor p53, which regulates the expression of genes involved in cell cycle arrest or apoptosis in response to genotoxicity or other cell stress, was significantly down-regulated in hypertrophic scar.[39] p53 can bind to the matrix metalloproteinase 2 (MMP-2) promoter and up-regulate its activity, thus decreasing net deposition of collagen by fibroblasts.

More recently direct evidence regarding delayed apoptosis was reported by Moulin *et al.* who examined anti-apoptotic and pro-apoptotic gene expression.[37] Isolated fibroblasts from normal healing wounds showed higher rates of apoptosis than those from hypertrophic scar; meanwhile, low levels of the anti-apoptotic proteins Bcl-2 and $Bclx_L$ were detected in normal wound myofibroblasts. These cells showed an increase in the level of the pro-apoptotic Bax when compared to normal skin fibroblasts or to myofibroblasts from hypertrophic scar. Myofibroblasts from hypertrophic scar showed a higher level of Bcl-2 compared to fibroblasts but no difference in the Bax or $Bclx_L$ levels. After serum starvation, wound myofibroblasts showed an increased apoptotic rate; however, hypertrophic scar myofibroblasts and fibroblasts did not show any difference. Anti-Fas treatment did not modify the levels of apoptosis but strongly increased the growth of hypertrophic scar myofibroblasts as compared to normal wound myofibroblasts. More recently, Linge *et al.* observed that hypertrophic scar fibroblasts exhibit resistance to a specific form of apoptosis elicited by contraction of collagen lattices, and that this phenomenon is dependent on the excess activity of cell surface tissue transglutaminase.[40] Taken together, these data support the hypothesis of defects in apoptosis and growth during pathological scar formation, resulting in delayed disappearance of fibroblasts and myofibroblasts.

Increased Levels of the Profibrotic Growth Factors TGF-β and CTGF in Hypertrophic Scar

TGF-β1 is the prototypic member of a structurally related protein family that is involved in many proliferative, inductive and regulatory processes.[41] In mammals, there are three TGF-β isoforms (β1, β2, and β3). The action of TGF-β1 has been characterized more extensively than that of the other two members. Signal transduction in response to TGF-β is initiated following ligand binding to heteromeric complexes of high-affinity cell surface receptors. Intracellular responses are mediated by a set of second messengers known as Smads.[41]

In wound healing, TGF-β1 influences the inflammatory response, angiogenesis, re-epithelialization, extracellular matrix deposition and remodeling.[42] TGF-β1 is one of the first cytokines to elicit inflammatory cell recruitment. The inflammatory phase is thought to be initiated by the release of TGF-β1 and other growth factors from the α granules of degranulating platelets. TGF-β1 is chemotactic and mitogenic. After migrating to the wound site, inflammatory cells synthesize and secrete additional TGF-β1, which at higher concentrations may induce the expression of other growth factors, thereby increasing the cellularity of the wound. In the maturation phase of healing, TGF-β1 continues to exert control over extracellular matrix components such as collagen and proteoglycan, not only by promoting their production, but also by reducing the expression of matrix metalloproteinases that would otherwise serve to break them down. In addition to its role in tissue repair, TGF-β1 is a potent regulator of the immune response.

TGF-β is one of the most prominent profibrotic growth factors in wound healing.[5] Interestingly, the expression profile of this growth factor in adult wounds is different from that of the fetus.[43] TGF-β1 is upregulated in all animal models of adult wound healing. However, TGF-β1 is barely detectable in fetal wounds.[43] TGF-β1 is overexpressed in hypertrophic scar.[3,4] Exogenous application of TGF-β1 to fetal wounds has been shown to result in the fetal response becoming adult-like, with fibroblast proliferation and collagen accumulation.[44]

Moreover, experimental manipulation of the wound environment in adult mice, rats and pigs, to mimic this growth factor profile, also resulted in scarless healing.[45] Transfection of a truncated TGF-β RII gene down-regulated TGF-β1 expression in rat incisional wounds and resulted in decreased inflammatory infiltrate, faster re-epithelialization, and less scarring,[42] indicating a central role of TGF-β1 in wound healing and scar formation.

Another growth factor recently implicated in pathological scar formation is connective tissue growth factor (CTGF). Previous studies have indicated that the stimulation of extracellular matrix synthesis by TGF-β is not shared by other growth factors such as fibroblast growth factor (FGF), epidermal growth factor (EGF), and platelet derived growth factor (PDGF). Recent evidence has demonstrated that CTGF is a downstream mediator of many of the effects of TGF-β on fibroblasts.[46]

CTGF is a cysteine-rich, heparin-binding protein with a molecular weight of 38 kDa, first identified in conditioned medium from human umbilical vein endothelial cells. It has a variety of effects on fibroblasts, including stimulating mitosis, chemotaxis and proliferation.[47] It was shown to directly stimulate fibroblast DNA synthesis and to augment the activity of other growth factors such as bFGF.[48] CTGF also stimulates the production of ECM components, such as collagen I, fibronectin and α5-integrin by fibroblasts.[49]

CTGF is also present and overexpressed in fibrotic skin disorders such as systemic sclerosis, localized skin sclerosis, keloids and scar tissue.[50] Studies of diseased tissues from human clinical specimens or animal models have shown that CTGF is overexpressed in fibrotic lesions of major organs and tissues, including atherosclerosis, renal diseases, hepatic fibrosis and malignant melanoma.[51] A common finding in these diseases is high-level expression of CTGF in fibroproliferative areas of affected tissues (50–100 fold compared to normal tissue).[49] In all cases, a direct correlation between high levels of CTGF expression and high levels of TGF-β expression could be established.[51]

In a pig skin wound healing model, CTGF expression peaked later[31] and subsequent to peak expression of TGF-β.[52] The coordinate expression of the two growth factors was interpreted as a component of a growth factor cascade in which TGF-β initiated regeneration and repairs, and stimulated the production of CTGF that was required later in wound healing. Many of the profibrogenic properties of TGF-β may be due to the induction of CTGF, which then stimulates fibroblast proliferation and ECM production.[49]

Connective tissue growth factor mRNA levels were reported to be 20-fold higher in fibroblasts from hypertrophic scar as compared to normal fibroblasts. When stimulated with TGF-β1, fibroblasts from hypertrophic scar showed a greater than 150-fold increase in CTGF mRNA expression compared with normal fibroblasts.[53] Not only TGF-β1 but also TGF-β2 and 3, could stimulate hypertrophic scar fibroblasts to increase CTGF mRNA expression over 100-fold compared with normal fibroblasts. Hypertrophic scar fibroblasts have both intrinsic up-regulation of CTGF transcription and an exaggerated capacity for CTGF transcription in response to stimulation by TGF-β. These data further indicate that both TGF-β and CTGF play roles in pathological scar formation.[53]

Hypertrophic Scarring is Associated with Blood Borne Fibrocytes

It is commonly accepted that dermal fibroblasts migrate into the injury site from the surrounding tissue when healing a small wound. When extensive areas of the skin are burned, it may be difficult or impossible for fibroblasts to migrate from the edges of the uninjured tissue, yet healing of extensive burn wounds often leads to excessive deposition of extracellular matrix in the dermis and the development of hypertrophic scarring.[3,5] It is not known from where and how these cells are recruited into this type of wound.

Recently, an emerging cell type, the fibrocyte, seems to be associated with post-burn hypertrophic scar formation.[54,55] Fibrocytes are a newly identified cell population initially discovered by their rapid and specific recruitment from blood to implanted wound chambers in mice.[56] They constitute 0.1~0.5% of peripheral blood cells and exhibit both monocyte and fibroblast-like characteristics.[57]

Fibrocytes are characterized by the expression of collagen type I, fibronectin, CD11b, CD34 and CD45, but not CD14, CD3 or CD10. They are found in the peripheral blood, wound sites and areas of tissue remodeling.[57] Fibrocytes originate from the bone marrow and have been reported also to contribute to the myofibroblast population in wounds.[58] These cells are thought to play a role in tissue repair by several mechanisms such as secretion of ECM, antigen presentation, cytokine production, angiogenesis and wound closure.[56,59–61] For example, fibrocytes have been shown to rapidly enter sites of tissue injury and contribute to wound healing by producing ECM macromolecules such as collagen type I, collagen type III and fibronectin.[56,59] Fibrocytes are potent instigators of the immune response by presenting antigen to both $CD4^+$ and $CD8^+$ T-lymphocytes and by secreting chemoattractants.[62,63] Fibrocytes have been reported to induce angiogenesis both *in vitro* and *in vivo*.[61] These cells may also provide a contractile force for wound closure via αSMA expression.[59] In addition, fibrocytes purified from wound chambers were found to express mRNA for IL-1β, IL-6, TNF-α,[64] and pro-inflammatory cytokines, which could prolong the lifespan of PMNs by delaying apoptosis. Recent work in our laboratory has established a link between fibrocytes and hypertrophic scar formation.[54,55]

Increased Numbers of Fibrocytes can be Cultured from the Blood of Burn Patients

We initially hypothesized that circulating fibrocytes might represent an important source of fibroblasts for healing of extensive burn wounds, where it may be difficult for fibroblasts to migrate from the edges of uninjured tissue. We therefore identified and quantified fibrocytes among the adherent cells cultured from human peripheral blood mononuclear cells (PBMC) obtained from 18 burn patients and 12 normal individuals, based on their ability to express type I collagen.[54] Our studies showed that adherent cells cultured from PBMC of burn patients developed into fibrocytes more efficiently than did those from normal individuals: the percentage of type I collagen-positive fibrocytes was significantly higher for burn patients than for controls ($89.7 \pm 7.9\%$ versus $69.9 \pm 14.7\%, p < 0.001$). Moreover, this percentage was consistently higher for patients with a more than 30% total body surface area burn up to one year, with the highest percentage appearing within three weeks of injury. A positive correlation was found between the levels of serum, TGF-β1 and the percentage of fibrocytes developing in the cultures of PBMC derived from these patients. We also demonstrated that fibrocytes were derived from CD14+ cells but not CD14− cells. Conditioned medium from CD14− cells was, however, required for fibrocyte differentiation, whereas direct contact between CD14− and CD14+ cells was not necessary. Treatment of the cell cultures with TGF-β1 enhanced the development of collagen-positive cells, whereas the inclusion of neutralizing anti-TGF-β1 antibodies in the CD14− conditioned medium suppressed fibrocyte differentiation.[54] These data suggest that the development of fibrocytes is up-regulated systemically in burn patients. Increased TGF-β in serum stimulates the differentiation of the CD14+ cell population in PBMC into collagen-producing cells that may be important in wound healing and scarring.

Establishment of LSP-1 as a Fibrocyte Marker

Despite sharing characteristics such as a spindle shape and the ability to synthesize collagen and other extracellular matrix macromolecules, fibrocytes and fibroblasts are different cell types.[55] While fibrocytes and fibroblasts in culture or in tissue appear in the light microscope to be spindle-shaped, when examined in the scanning electron microscope they exhibit distinct morphologies (Fig. 2A). When PBMCs are cultured, some initially round cells attach and polarize. From day 7 in culture, their morphology appears to be stable and cell size variation is the only significant change. Mature fibrocytes are characterized by an elongated body with a cluster of fiber-like projections at one end. The surface of the cell body has numerous blebs. In contrast, fibroblasts in culture show a mainly smooth cell body with relatively well-defined lamellipodia and ruffled borders,

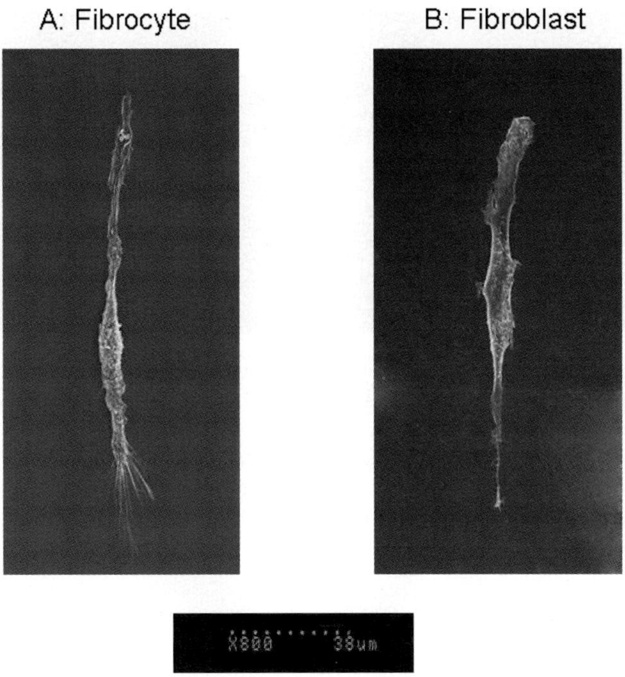

Fig. 2. Scanning electron microscopy images of a mature fibrocyte after 14 days in culture (Panel A) and a fibroblast (Panel B).

and a single retraction fiber (Fig. 2B). Fibrocytes are characterized by the expression of collagen type I, fibronectin, CD11b, CD34 and CD45 in the peripheral blood, wound sites and areas of tissue remodeling.[57] Unfortunately, these cell surface proteins are not unique to fibrocytes: CD34, for example, is also found on capillary endothelial cells. Moreover, some of these markers, including CD34 and CD45, are gradually lost in culture (Wang *et al.*, unpublished).[65] To seek a more stable marker for fibrocytes, we extracted proteins from fibroblasts, fibrocytes and non-adherent lymphocytes and examined them by 2-dimensional gel electrophoresis and mass spectrometry.[55] We found that leukocyte specific protein 1 (LSP-1) is present at higher levels in fibrocytes from burn patients than in non-adherent lymphocytes or in fibrocytes from normal subjects. LSP-1 is completely absent from fibroblasts. Dual immunostaining for procollagen I and

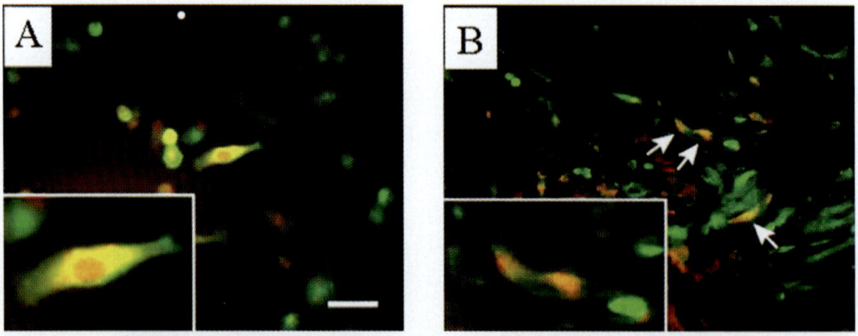

Fig. 3. Dual immunofluorescent staining of fibrocytes in culture (Panel A) and in the dermis of hypertrophic scar tissue (Panel B). Cells cultured on coverslips and cryosections of hypertrophic scar were stained with antibodies specific for human LSP-1 (green) and for the N-terminal propeptide of type I collagen (red). Note that fibrocytes are labeled by both antibodies and exhibit a yellow color located predominantly around the nucleus (Scale bar 100μm). (From Yang *et al.*, 2005, with permission.)

LSP-1 can therefore be used to identify fibrocytes (and to distinguish them from fibroblasts), both in culture and in hypertrophic scars (Fig. 3).

Leukocyte-specific protein 1 (formerly known as lymphocyte-specific protein 1 but renamed to reflect its expression in macrophages and neutrophils) is a 52-kDa intracellular F-actin binding protein that accumulates on the cortical cytoskeleton. LSP-1 is expressed in mature and immature B and T cells, macrophages and neutrophils.[66] Human and mouse LSP-1 share the same expression patterns and their level of sequence identity is high. LSP-1 has two putative Ca^{2+} binding motifs and distributes in three different subcellular fractions: ~25% is in the cytoplasmic face of the plasma membrane; ~15% in the cytoskeleton; and ~60% in the cytosol.[67] LSP-1 is a substrate for mitogen-activated protein kinase activated protein kinase 2 and for protein kinase C,[66] two enzymes implicated in leukocyte migration and chemotaxis. In addition, LSP-1 was also reported to regulate a Ca^{2+}-dependent step in the induction of anti-IgM mediated apoptosis.[68] Therefore, LSP-1 appears to be a signaling molecule regulating cytoskeletal architecture and mobility and playing a role in receptor-induced apoptosis. In hairy cell leukemia, overexpression of LSP-1 inhibits neutrophil migration and increases cell adhesion, leading to recurrent infection.[69] Therefore, LSP-1 might regulate fibrocyte motility. Alternatively, the activation of a lymphocyte subpopulation (fibrocyte precursors) might cause expression of LSP-1 and induce adhesion and differentiation into fibrocytes. A previous study found that fibrocytes bind tightly to plastic or glass surfaces and proliferate only slowly in culture.[54] If the same is true *in vivo*, increased binding of fibrocytes to the newly formed matrix might inhibit their migration within it and affect the deposition and orientation of collagen. Our preliminary study demonstrated that LSP-1 is expressed by fibrocytes in culture for at least seven weeks, while CD34 gradually disappears (Wang *et al.*, unpublished data).

Increased Numbers of Fibrocytes in Post-burn Hypertrophic Scar

Fibrocytes can be distinguished from fibroblasts using staining for LSP-1 in combination with type I procollagen. This was used to demonstrate that there were more fibrocytes in post-burn hypertrophic scar than in mature scar.[55] Both hypertrophic scar and mature scar tissue showed extensive intracellular staining for type I procollagen. Some cells with cortical cytoplasmic staining for LSP-1 were also seen in the dermis. In double-exposure photographs, fibrocytes are apparent as dual-labeled spindle-shaped cells (Fig. 3B). The greater number of fibrocytes in hypertrophic tissue than in mature scar was confirmed by counting of the number of dual-labeled cells per high-power field (2.4% ± 0.5% vs. 1.4% ± 0.5%, $p < 0.05$, $n = 5$). A few procollagen-stained fibroblasts were seen in the dermis in normal skin, but fibrocytes were not detected.

More recently, we have used flow cytometry to quantify fibrocytes. Skin and scar tissue were treated with dispase to separate the dermis from the epidermis, the single cell suspension was labeled and flow cytometry analysis was performed. Twenty samples of normal skin, 15 of mature scar and 9 of hypertrophic scar, were analyzed (Fig. 4). Normal dermis presents a fibrocyte population of 0.45%, which is comparable to that in mature scar. However, hypertrophic scars had on average 0.81% fibrocytes.

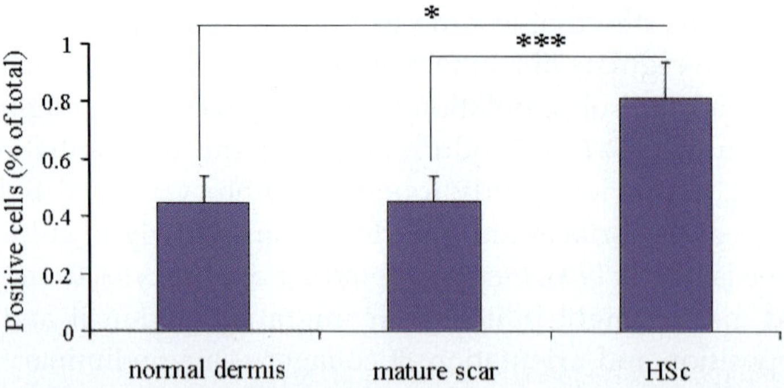

Fig. 4. Quantification of fibrocytes by flow cytometry in skin tissues. Dermal sections from normal skin samples ($n = 20$), mature scars ($n = 15$) and hypertrophic scars ($n = 9$) were subjected to enzyme digestion; cell suspensions were prepared, labeled with antibodies specific for human LSP-1 and for the N-terminal propeptide of type I collagen and subsequently analyzed by FACS. Note the higher percentage of fibrocytes present in hypertrophic scar compared to normal skin or mature scar. (* PL0.05, *** PL0.001)

Potential Interaction of Fibrocytes and Endothelial Cells

Fibrocytes not only can produce collagen I, fibronectin and vimentin and contract collagen lattice gels, thus directly contributing to wound healing,[56,59] but they can also interact with other cells in the dermis. They may therefore play an indirect role in healing and in abnormal scarring. Fibrocytes have been reported to be able to induce an angiogenic phenotype in microvascular endothelial cells *in vitro* and to promote angiogenesis *in vivo*.[61] They constitutively secrete extracellular matrix-degrading enzymes, primarily MMP 9, which promotes endothelial cell invasion. In addition, fibrocytes secrete several proangiogenic factors, including VEGF, bFGF, IL-8, PDGF and hematopoietic growth factors that promote endothelial cell migration, proliferation, and/or tube formation. By contrast, they do not produce representative antiangiogenic factors. Finally, both autologous fibrocytes and fibrocyte-conditioned media were found to induce blood vessel formation *in vivo* using the Matrigel angiogenesis model.[61]

Possible Interactions of Fibrocytes and Fibroblasts

Since dermal fibroblasts sit in a pivotal position in wound healing and abnormal scar formation, we have recently examined the possible role of fibrocytes in regulating the activities of fibroblasts. Compared to fibroblasts, fibrocytes produce very small amounts of collagen *in vitro* (as measured by hydroxyproline in the medium). However, conditioned medium from cultured burn patient fibrocytes, but not that from normal subjects' fibrocytes, increases hydroxyproline synthesis by dermal fibroblasts. This medium also stimulated dermal fibroblasts to proliferate and migrate, to express αSMA and to differentiate into myofibroblasts, and to contract fibroblast-populated collagen lattices (Wang *et al.*, submitted for publication). Our findings, therefore, indicate a potential regulatory role of fibrocytes in post-burn pathological scar formation.

Elevated TGF-β and CTGF mRNA Levels in Burn Patient Fibrocytes

We recently examined the levels of TGF-β1 in fibrocyte-conditioned medium from normal subjects and burn patients using ELISA. We found that normal subjects' fibrocytes produced about 18 pg per 1000 cells, but more than twice this amount of TGF-β1 was found in conditioned medium derived from burn patient fibrocytes (42.6 ± 5.3 vs. 17.9 ± 2.1, $p < 0.01$) (Fig. 5A). A neutralizing antibody to TGF-β1 significantly reduced the stimulation of dermal fibroblast proliferation and collagen lattice gel contraction by burn patient fibrocyte-conditioned medium. Therefore, TGF-β1 is a key factor produced by burn patient fibrocytes. Using RT-PCR, we have also found significant increases in mRNA levels for CTGF in burn patient fibrocytes compared to fibrocytes from normal subjects (Fig. 5B). As TGF-β1 and CTGF have been reported to coordinately regulate wound healing, it is likely that these two growth factors from burn patient fibrocytes also function together in regulating dermal fibroblast activity following thermal injury.

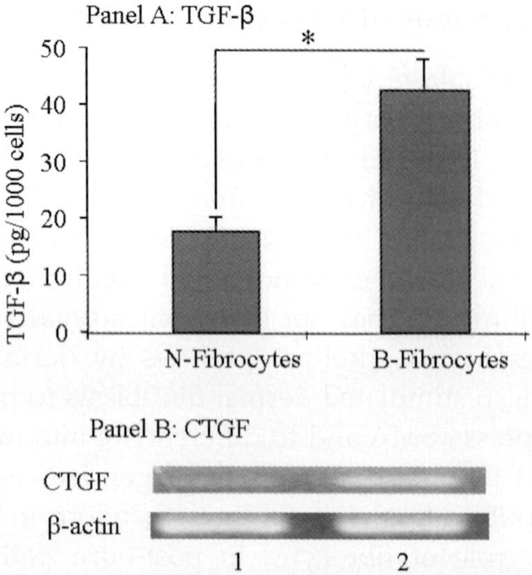

Fig. 5. Production of TGF-β and CTGF by fibrocytes. Panel A: TGF-β production. The concentration of TGF-β was measured by ELISA in medium conditioned by the culture of purified fibrocytes from normal subjects ("N-Fibrocytes"), or from burn patients ("B-Fibrocytes"). Panel B: Comparison of levels of mRNA for CTGF. Total RNA was extracted from purified fibrocytes, reverse-transcribed and the cDNAs for CTGF and β-actin were amplified by PCR. Lane 1: fibrocytes from normal subjects; Lane 2: fibrocytes from burn patients. The * indicates a significant difference ($P < 0.015$).

Fibrocytes may Contribute to the Myofibroblast Population

It has been reported that cultured fibrocytes are able to differentiate into myofibroblasts, cells that are characterized by the expression of αSMA and the ability to contract a collagen lattice in the presence of TGF-β.[59] Fibrocytes may also differentiate into myofibro-blasts in healing wounds in BALB/c mice.[58] During wound healing, there was a marked increase in the number of cells expressing αSMA in the granulation tissue. Between 4 and 7 days after wounding, more than 50% of these cells also expressed the CD13 antigen. CD13+/collagen I+ fibrocytes could be isolated by fluorescence-activated cell sorting at an early stage of healing from digested fragments of wounded tissue. Like authentic fibrocytes, these cells were also CD45+/CD34+/CD14−. Between 4 and 7 days post-injury, 61.4% of the isolated fibrocytes were found to express αSMA. This recent study shows that fibrocytes can contribute to the myofibroblast population *in vivo*. In a bleomycin induced murine pulmonary fibrosis study, more fibrocytes were found in the lung. These fibrocytes can also apparently differentiate into myofibroblasts.[65] In addition, our recent work shows that fibrocytes from burn patient can stimulate dermal fibroblasts to differentiate into functional myofibroblasts. Taken together, these data implicate fibrocyte-derived myofibroblasts in hypertrophic scar formation.

Possible Role of Fibrocytes in the Polarized Th2 Immune Response

Our recent longitudinal work shows a polarized Th2 immune response in post-burn hypertrophic scarring,[15] indicating that cytokines secreted by different T-cell subsets and other immune cells may have a role in the development of hypertrophic scar. Recently, Grab *et al.* suggested that fibrocytes might play a role in the polarized Th2 immune response in leishmaniasis.[70] Fibrocytes appear to be an abundant source of cytokines, chemoattractants and growth factors. In addition, fibrocytes are able to recruit and activate naive T cells and memory T cells.[56,62,64,71] Using a human inflammatory cytokine microarray, we have demonstrated that fibrocytes can express chemotactic cytokines such as monocyte chemotactic protein 1 (MCP-1), macrophage inflammatory protein 1α (MIP-1α) and MIP-1β; pro-inflammatory cytokines such as IL-6 and tumor necrosis factor α (TNF-α); and Th2-representative cytokines such as IL-10 (Tredget *et al.*, unpublished). It is known that the chemotactic cytokines mediate the recruitment of inflammatory cells, and that the proinflammatory cytokines contribute to the host immunological response. Since fibrocytes express high levels of B7–2 (CD86),[62,71] they may also have a role in the signaling of T cells that mediates Th2 immune responses.[72] In addition, from our inflammatory cytokine microarray results, IFN-γ and IL-2 (Th1 cytokines) mRNA levels in burn patient fibrocytes are down-regulated compared to those of normal subjects. Meanwhile, TGF-β protein levels in conditioned medium from burn patient fibrocytes were significantly upregulated (Fig. 5A). The phenotype of high levels of B7–2 expression and low IFN-γ production, suggests that fibrocytes may have the potential to influence the Th2 response in burn patients. The expression of IL-6 and IL-10, together with TGF-β1, could further skew an emerging Th1 response toward the Th2 phenotype. Therefore, fibrocytes might play a role in the directing the immune response away from a protective Th1 response towards a pathogenic Th2 response in hypertrophic scarring.

Proposed Role of Fibrocytes in Hypertrophic Scar Formation

Circulating fibrocytes infiltrating into injured tissue during wound healing may not only produce extracellular matrix (ECM) molecules but also regulate the functions of the surrounding cells such as endothelial cells and fibroblasts (Fig. 6). Thermal injury activates

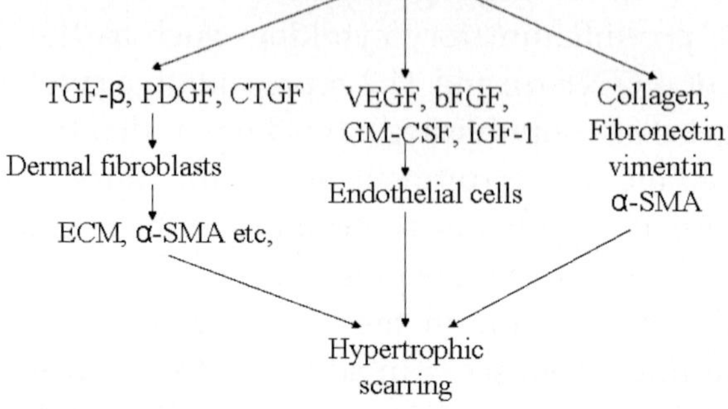

Fig. 6. Proposed role of fibrocytes in hypertrophic scar formation.

progenitor fibrocytes and the interaction between fibrocytes and activated T cells promotes their early differentiation. Following differentiation, fibrocytes migrate into the injured sites where they further differentiate and contribute to ECM production, angiogenesis and wound contraction. In addition, fibrocytes produce profibrotic factors such as TGF-β1 and CTGF to activate local fibroblasts, ultimately amplifying the wound healing process.

Summary and Prospects for Future Work

Wound healing requires an elaborate interplay between several cell types that orchestrate a series of regulated and overlapping events. Fibrocytes are a unique leukocyte subpopulation implicated in this process. Although the pathogenesis of hypertrophic scar formation following thermal injury is still incompletely understood, investigations into the composition of the tissue itself, the activities of scar fibroblasts, effects of various cytokines and growth factors, and the impact of the immune response have all contributed to understanding this disorder. The recent discovery of the fibrocyte has opened up new avenues of investigation that may lead eventually to novel treatments for the prevention or treatment of post-burn hypertrophic scarring. Before this is possible, however, it will be necessary for us to learn more about what controls the recruitment, differentiation and activation of these cells in burn wounds and what would be the consequences of eliminating them. These questions are not easily addressed in human subjects and their study will almost certainly require experimental animal models.

Acknowledgments

We thank Haiyan Jiao, Tara Lynn Stewart, and Heather Shankowsky for their contributions to the original work described in this chapter.

References

1. Colwell AS, Longaker MT, Lorenz HP. (2003) Fetal wound healing. *Front Biosci* **8**: s1240–1248.

2. Tredget EE. (1994) The molecular biology of fibroproliferative disorders of the skin: Potential cytokine therapeutics. *Ann Plast Surg* **33**: 152–154.
3. Tredget EE, Nedelec B, Scott PG, Ghahary A. (1997) Hypertrophic scars, keloids, and contractures. The cellular and molecular basis for therapy. *Surg Clin North Am* **77**: 701–730.
4. Tredget EE. (1999) Pathophysiology and treatment of fibroproliferative disorders following thermal injury. *Ann N Y Acad Sci* **888**: 165–182.
5. Scott PG, Ghahary A, Tredget EE. (2000) Molecular and cellular aspects of fibrosis following thermal injury. *Hand Clin* **16**: 271–287.
6. Scott PG, Ghahary A, Chambers MM, Tredget EE. (1994) Biological basis of hypertrophic scarring. *Advances in Structural Biology* **3**: 157–202.
7. Scott PG, Dodd CM, Tredget EE, *et al.* (1996) Chemical characterization and quantification of proteoglycans in human post-burn hypertrophic and mature scars. *Clin Sci (Lond)* **90**: 417–425.
8. Linares HA, Kischer CW, Dobrkovsky M, Larson DL. (1972) The histiotypic organization of the hypertrophic scar in humans. *J Invest Dermatol* **59**: 323–331.
9. Alexander SA, Donoff RB. (1980) The histochemistry of glycosaminoglycans within hypertrophic scars. *J Surg Res* **28**: 171–181.
10. Scott PG, Dodd CM, Tredget EE, *et al.* (1995) Immunohistochemical localization of the proteoglycans decorin, biglycan and versican and transforming growth factor-beta in human post-burn hypertrophic and mature scars. *Histopathology* **26**: 423–431.
11. Scott PG, Dodd CM, Ghahary A, *et al.* (1998) Fibroblasts from post-burn hypertrophic scar tissue synthesize less decorin than normal dermal fibroblasts. *Clin Sci (Lond)* **94**: 541–547.
12. Harty M, Neff AW, King MW, Mescher AL. (2003) Regeneration or scarring: An immunologic perspective. *Dev Dyn* **226**: 268–279.
13. Castagnoli C, Trombotto C, Ondei S, *et al.* (1997) Characterization of T-cell subsets infiltrating post-burn hypertrophic scar tissues. *Burns* **23**: 565–572.
14. Castagnoli C, Stella M, Magliacani G. (2002) Role of T-lymphocytes and cytokines in post-burn hypertrophic scars. *Wound Repair Regen* **10**: 107–108.
15. Tredget EE, Yang L, Delehanty M, *et al.* (2006) Polarized Th2 cytokine production in patients with hypertrophic scar following thermal injury. *J Interferon Cytokine Res* **26**: 179–189.

16. Mosmann TR, Coffman RL. (1989) TH1 and TH2 cells: Different patterns of lymphokine secretion lead to different functional properties. *Annu Rev Immunol* **7**: 145–173.
17. Mosmann TR. (1991) Role of a new cytokine, interleukin-10, in the cross-regulation of T helper cells. *Ann N Y Acad Sci* **628**: 337–344.
18. Peng JK, Lin JS, Kung JT, *et al.* (2005) The combined effect of IL-4 and IL-10 suppresses the generation of, but does not change the polarity of, type-1 T cells in Histoplasma infection. *Int Immunol* **17**: 193–205.
19. Hoffmann KF, Cheever AW, Wynn TA. (2000) IL-10 and the dangers of immune polarization: Excessive type 1 and type 2 cytokine responses induce distinct forms of lethal immunopathology in murine schistosomiasis. *J Immunol* **164**: 6406–6416.
20. Wynn TA. (2004) Fibrotic disease and the T(H)1/T(H)2 paradigm. *Nat Rev Immunol* **4**: 583–594.
21. Caligaris-Cappio F, Bertero MT, Converso M, *et al.* (1995) Circulating levels of soluble CD30, a marker of cells producing Th2-type cytokines, are increased in patients with systemic lupus erythematosus and correlate with disease activity. *Clin Exp Rheumatol* **13**: 339–343.
22. Holmes A, Abraham DJ, Chen Y, *et al.* (2003) Constitutive connective tissue growth factor expression in scleroderma fibroblasts is dependent on Sp1. *J Biol Chem* **278**: 41728–41733.
23. Matsushima H, Hayashi S, Shimada S. (2003) Skin scratching switches immune responses from Th2 to Th1 type in epicutaneously immunized mice. *J Dermatol Sci* **32**: 223–230.
24. Horgan AF, Mendez MV, O'Riordain DS, *et al.* (1994) Altered gene transcription after burn injury results in depressed T-lymphocyte activation. *Ann Surg* **220**: 342–351; discussion 351–342.
25. O'Sullivan ST, Lederer JA, Horgan AF, *et al.* (1995) Major injury leads to predominance of the T helper-2 lymphocyte phenotype and diminished interleukin-12 production associated with decreased resistance to infection. *Ann Surg* **222**: 482–490; discussion 490–482.
26. Desmouliere A, Badid C, Bochaton-Piallat ML, Gabbiani G. (1997) Apoptosis during wound healing, fibrocontractive diseases and vascular wall injury. *Int J Biochem Cell Biol* **29**: 19–30.
27. Greenhalgh DG. (1998) The role of apoptosis in wound healing. *Int J Biochem Cell Biol* **30**: 1019–1030.

28. Martin P. (1997) Wound healing — aiming for perfect skin regeneration. *Science* **276**: 75–81.
29. Savill JS, Wyllie AH, Henson JE, *et al.* (1989) Macrophage phagocytosis of aging neutrophils in inflammation. Programmed cell death in the neutrophil leads to its recognition by macrophages. *J Clin Invest* **83**: 865–875.
30. Witko-Sarsat V, Rieu P, Descamps-Latscha B, *et al.* (2000) Neutrophils: Molecules, functions and pathophysiological aspects. *Lab Invest* **80**: 617–653.
31. Wang JF, Olson ME, Reno CR, *et al.* (2001) The pig as a model for excisional skin wound healing: Characterization of the molecular and cellular biology, and bacteriology of the healing process. *Comp Med* **51**: 341–348.
32. Meszaros AJ, Reichner JS, Albina JE. (1999) Macrophage phagocytosis of wound neutrophils. *J Leukoc Biol* **65**: 35–42.
33. Deitch EA, Wheelahan TM, Rose MP, *et al.* (1983) Hypertrophic burn scars: Analysis of variables. *J Trauma* **23**: 895–898.
34. Desmouliere A, Chaponnier C, Gabbiani G. (2005) Tissue repair, contraction, and the myofibroblast. *Wound Repair Regen* **13**: 7–12.
35. Nedelec B, Dodd CM, Scott PG, *et al.* (1998) Effect of interferon-alpha2b on guinea pig wound closure and the expression of cytoskeletal proteins *in vivo*. *Wound Repair Regen* **6**: 202–212.
36. Nedelec B, Ghahary A, Scott PG, Tredget EE. (2000) Control of wound contraction. Basic and clinical features. *Hand Clin* **16**: 289–302.
37. Moulin V, Larochelle S, Langlois C, *et al.* (2004) Normal skin wound and hypertrophic scar myofibroblasts have differential responses to apoptotic inductors. *J Cell Physiol* **198**: 350–358.
38. Wassermann RJ, Polo M, Smith P, *et al.* (1998) Differential production of apoptosis-modulating proteins in patients with hypertrophic burn scar. *J Surg Res* **75**: 74–80.
39. Sheikh MS, Fornace AJ, Jr. (2000) Role of p53 family members in apoptosis. *J Cell Physiol* **182**: 171–181.
40. Linge C, Richardson J, Vigor C, *et al.* (2005) Hypertrophic scar cells fail to undergo a form of apoptosis specific to contractile collagen — the role of tissue transglutaminase. *J Invest Dermatol* **125**: 72–82.
41. Leask A, Abraham DJ. (2004) TGF-beta signaling and the fibrotic response. *Faseb J* **18**: 816–827.

42. Liu W, Wang DR, Cao YL. (2004) TGF-beta: A fibrotic factor in wound scarring and a potential target for anti-scarring gene therapy. *Curr Gene Ther* **4**: 123–136.
43. Longaker MT, Peled ZM, Chang J, Krummel TM. (2001) Fetal wound healing: Progress report and future directions. *Surgery* **130**: 785–787.
44. Krummel TM, Michna BA, Thomas BL, *et al.* (1988) Transforming growth factor beta (TGF-beta) induces fibrosis in a fetal wound model. *J Pediatr Surg* **23**: 647–652.
45. Ferguson MW, O'Kane S. (2004) Scar-free healing: From embryonic mechanisms to adult therapeutic intervention. *Philos Trans R Soc Lond B Biol Sci* **359**: 839–850.
46. Grotendorst GR. (1997) Connective tissue growth factor: A mediator of TGF-beta action on fibroblasts. *Cytokine Growth Factor Rev* **8**: 171–179.
47. Frazier K, Williams S, Kothapalli D, *et al.* (1996) Stimulation of fibroblast cell growth, matrix production, and granulation tissue formation by connective tissue growth factor. *J Invest Dermatol* **107**: 404–411.
48. Wang JF, Olson ME, Ball DK, *et al.* (2003) Recombinant connective tissue growth factor modulates porcine skin fibroblast gene expression. *Wound Repair Regen* **11**: 220–229.
49. Brigstock DR. (2003) The CCN family: A new stimulus package. *J Endocrinol* **178**: 169–175.
50. Igarashi A, Nashiro K, Kikuchi K, *et al.* (1996) Connective tissue growth factor gene expression in tissue sections from localized scleroderma, keloid, and other fibrotic skin disorders. *J Invest Dermatol* **106**: 729–733.
51. Moussad EE, Brigstock DR. (2000) Connective tissue growth factor: What's in a name? *Mol Genet Metab* **71**: 276–292.
52. Wang JF, Olson ME, Reno CR, *et al.* (2000) Molecular and cell biology of skin wound healing in a pig model. *Connect Tissue Res* **41**: 195–211.
53. Colwell AS, Phan TT, Kong W, *et al.* (2005) Hypertrophic scar fibroblasts have increased connective tissue growth factor expression after transforming growth factor-beta stimulation. *Plast Reconstr Surg* **116**: 1387–1390; discussion 1391–1382.
54. Yang L, Scott PG, Giuffre J, *et al.* (2002) Peripheral blood fibrocytes from burn patients: Identification and quantification of fibrocytes in adherent cells cultured from peripheral blood mononuclear cells. *Lab Invest* **82**: 1183–1192.
55. Yang L, Scott PG, Dodd C, *et al.* (2005) Identification of fibrocytes in postburn hypertrophic scar. *Wound Repair Regen* **13**: 398–404.

56. Bucala R, Spiegel LA, Chesney J, *et al.* (1994) Circulating fibrocytes define a new leukocyte subpopulation that mediates tissue repair. *Mol Med* **1**: 71–81.
57. Quan TE, Cowper S, Wu SP, *et al.* (2004) Circulating fibrocytes: Collagen-secreting cells of the peripheral blood. *Int J Biochem Cell Biol* **36**: 598–606.
58. Mori L, Bellini A, Stacey MA, *et al.* (2005) Fibrocytes contribute to the myofibroblast population in wounded skin and originate from the bone marrow. *Exp Cell Res* **304**: 81–90.
59. Abe R, Donnelly SC, Peng T, *et al.* (2001) Peripheral blood fibrocytes: Differentiation pathway and migration to wound sites. *J Immunol* **166**: 7556–7562.
60. Metz CN. (2003) Fibrocytes: A unique cell population implicated in wound healing. *Cell Mol Life Sci* **60**: 1342–1350.
61. Hartlapp I, Abe R, Saeed RW, *et al.* (2001) Fibrocytes induce an angiogenic phenotype in cultured endothelial cells and promote angiogenesis *in vivo*. *Faseb J* **15**: 2215–2224.
62. Chesney J, Bacher M, Bender A, Bucala R. (1997) The peripheral blood fibrocyte is a potent antigen-presenting cell capable of priming naive T cells *in situ*. *Proc Natl Acad Sci USA* **94**: 6307–6312.
63. Balmelli C, Ruggli N, McCullough K, Summerfield A. (2005) Fibrocytes are potent stimulators of anti-virus cytotoxic T cells. *J Leukoc Biol* **77**: 923–933.
64. Chesney J, Metz C, Stavitsky AB, *et al.* (1998) Regulated production of type I collagen and inflammatory cytokines by peripheral blood fibrocytes. *J Immunol* **160**: 419–425.
65. Phillips RJ, Burdick MD, Hong K, *et al.* (2004) Circulating fibrocytes traffic to the lungs in response to CXCL12 and mediate fibrosis. *J Clin Invest* **114**: 438–446.
66. Jongstra-Bilen J, Misener VL, Wang C, *et al.* (2000) LSP1 modulates leukocyte populations in resting and inflamed peritoneum. *Blood* **96**: 1827–1835.
67. Jongstra-Bilen J, Janmey PA, Hartwig JH, *et al.* (1992) The lymphocyte-specific protein LSP1 binds to F-actin and to the cytoskeleton through its COOH-terminal basic domain. *J Cell Biol* **118**: 1443–1453.
68. Jongstra-Bilen J, Wielowieyski A, Misener V, Jongstra J. (1999) LSP1 regulates anti-IgM induced apoptosis in WEHI-231 cells and normal immature B-cells. *Mol Immunol* **36**: 349–359.

69. Miyoshi EK, Stewart PL, Kincade PW, *et al.* (2001) Aberrant expression and localization of the cytoskeleton-binding pp52 (LSP1) protein in hairy cell leukemia. *Leuk Res* **25**: 57–67.
70. Grab DJ, Salem ML, Dumler JS, Bucala R. (2004) A role for the peripheral blood fibrocyte in leishmaniasis? *Trends Parasitol* **20**: 12.
71. Grab DJ, Salim M, Chesney J, *et al.* (2002) A role for peripheral blood fibrocytes in Lyme disease? *Med Hypotheses* **59**: 1–10.
72. Hofer MF, Jirapongsananuruk O, Trumble AE, Leung DY. (1998) Upregulation of B7.2, but not B7.1, on B cells from patients with allergic asthma. *J Allergy Clin Immunol* **101**: 96–102.

Chapter 6

Role in Asthmatic Lung Disease

Sabrina Mattoli*,†,‡ and Matthias Schmidt†

Asthma is a chronic inflammatory disease of the airways characterized by structural and functional alterations of the bronchial epithelium and remodeling of the normal bronchial architecture. Bronchial myofibroblasts are thought to play a crucial role in the pathogenesis of subepithelial fibrosis, which represents a prominent feature of the remodeling process. Although it has been postulated for many years that myofibroblasts derive from tissue fibroblasts, recent studies have indicated that bone marrow-derived fibrocytes may contribute to the bronchial myofibroblast population in asthma and may be responsible for the excessive collagen deposition below the epithelial basement membrane. Investigating how fibrocytes emerge in asthmatic airways and what their fate is, may uncover key mechanisms involved in the pathogenesis of

*Address for correspondence: Sabrina Mattoli, M.D., Ph.D., Avail GmbH, P.O. Box 110, CH-4003 Basel, Switzerland, Tel.: +41 61 262 3564; Fax: +41 61 262 3562. E-mail: smattoli@avail-research.com

†Avail Biomedical Research Institute.

‡Avail GmbH, Basel, Switzerland.

airway remodeling and may help clarify the functional role of the remodeling process in asthma.

Introduction

Asthma is a chronic disease affecting about 10% of the population worldwide. It is characterized by structural[1–4] and functional[5–12] abnormalities of the bronchial epithelium;[1–3] accumulation of inflammatory cells (activated T lymphocytes, eosinophils and mast cells) in the bronchial mucosa;[13–15] and remodeling of the airway tissue structure.[14,15] Recurrent episodes of wheezing and shortness of breath, bronchoconstriction upon exposure to a variety of innocuous agents, and a rapid decline in lung function over time represent the clinical expression of these bronchial structural and functional alterations. Repeated cycles of airway inflammation and repair, possibly linked to a primary functional defect of the asthmatic bronchial epithelium,[5–12] are considered to represent the driving force for airway remodeling. Because the remodeling process has been suspected to cause the irreversible decline in lung function,[16,17] it has become a major target for the development of new anti-asthma drugs.[18]

Peculiar aspects of airway tissue remodeling in asthma include the accumulation of fibroblasts and myofibroblasts below the epithelial basement membrane and the thickening of the *lamina reticularis* (Fig. 1).[17,19–22] Because the *lamina reticularis* contains collagens I, III and V, fibronectin and a tenascin-rich matrix,[20,23] its thickening in asthma has been referred to as "subepithelial fibrosis" (Fig. 1). The numerous myofibroblasts present in the bronchial subepithelial area of asthmatic patients[17,20] (Fig. 1) are thought to mediate the majority of the events contributing to subepithelial fibrosis because they represent an important source of collagens and other extracellular matrix molecules,[24] but their origin is still uncertain.

The main objective of this chapter is to review the data indicating that bone marrow-derived fibrocytes may function as precursors of bronchial fibroblasts and myofibroblasts in asthma and may play a crucial role in the genesis of subepithelial fibrosis.

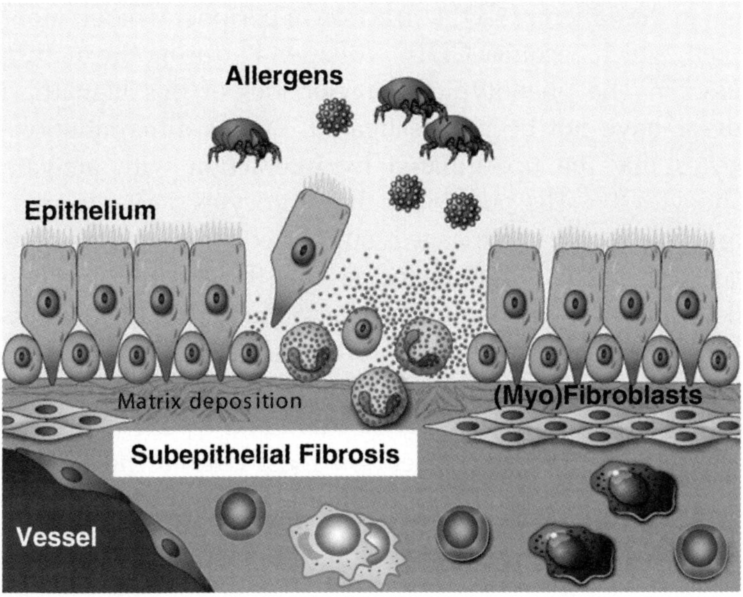

Fig. 1. The airway epithelium of asthmatic patients is frequently damaged and more susceptible than the epithelium of non-asthmatic individuals to environmental insults, such as exposure to allergens with proteolytic activity. Epithelial structural and functional alterations are associated with the accumulation of fibroblasts and myofibroblasts below the epithelial basement membrane and with excessive deposition of collagen and other extracellular matrix molecules in this area. The resulting thickening of the *lamina reticularis* is known as "subepithelial fibrosis."

Phenotypic and Functional Characteristics of Fibrocytes

Fibrocytes represent a unique population of cells that express fibroblast products, such as collagen I, in conjunction with the hematopoietic stem cell antigen CD34, the leukocyte common antigen CD45 and the markers of the myeloid lineage cells CD11b and CD13.[25–27] Although the bone marrow origin of fibrocytes has been established in an animal model of wound healing, using sex-mismatched chimera mice,[28] these cells are not normally present in the peripheral blood as such but emerge from cultures of peripheral blood mononuclear cells after 10–14 days.[25–27] They may originate from a $CD11b^+$ $CD13^+$ $CD45^+$ $CD34^+$ collagen I^- precursor present in the adherent $CD14^+$ fraction of peripheral blood mononuclear cell, which becomes $CD14^-$ collagen I^+ during the maturation process.[26,27] The phenotypic characteristics of this hematopoietic precursor have not been investigated, but its differentiation into fibrocytes may be upregulated by interaction with activated T lymphocytes.[26,27] The possibility that fibrocytes derive from a precursor of the monocyte lineage is supported by the results of recent *in vivo* experiments,[29] demonstrating that the selective depletion of circulating cells of the monocyte/macrophage lineage prevents the accumulation of fibrocytes in the pulmonary artery adventitia in animal models of hypoxia-induced pulmonary vascular remodeling.

The fact that fibrocytes can be identified by the expression of surface markers that are present on most cells of the monocyte/macrophage lineage should be taken into account when specimens of tissues or bone marrow are examined for the presence of fibrocytes by immunohistochemical or immunofluorescence analysis. Tissue monocytes and macrophages can in fact contain phagocytosed fragments of interstitial collagen I and can show intracellular staining after labeling with an antibody against collagen I. To avoid confusion, staining for collagen I should be performed by using antibodies against the collagen I precursor, procollagen I, or the expression of collagen I should be confirmed at the gene level, for example by *in situ* hybridization with a probe for procollagen I mRNA.[28,30]

The fibrocytes that emerge from cultures of peripheral blood mononuclear cells do not constitutively express the myofibroblast marker α-smooth muscle actin (αSMA),[26,30] but acquire the myofibroblast phenotype under *in vitro* stimulation with fibrogenic cytokines that are produced in exaggerated quantities in the airways of asthmatic patients, such as transforming growth factor (TGF-β_1)[31–33] and endothelin (ET-1).[34,35] When cultured in the presence of TGF-β_1 and ET-1 for 6 days, fibrocytes develop bundles of actin microfilaments indicative of a contractile phenotype and show the other ultrastructural characteristics of fibroblasts undergoing differentiation into myofibroblasts.[30] Stimulation with TGF-β_1 also increases collagen I immunoreactivity in cultured fibrocytes[26] and markedly enhances the release of collagens (type I and III) and fibronectin from these cells.[30] The differentiation of fibrocytes into fibroblasts and myofibroblasts *in vitro* is associated with a down-regulation of the expression of the surface antigens CD34,[28,30,36,37] CD45[28,37] and CD13.[28] Human and murine circulating fibrocytes are quite similar in terms of phenotypic characteristics and response to TGF-β_1 *in vitro*.[28,30]

ET-1 is known to induce the proliferation of bronchial fibroblasts[38] and smooth muscle cells[39] and this biological activity is one of the reasons why it has been implicated in the genesis of airway remodeling. Interestingly, ET-1 also greatly promotes the proliferation of cultured human fibrocytes in a concentration-dependent manner before they differentiate into myofibroblasts.[30]

Differentiation of Fibrocytes at the Tissue Sites

Two studies[28,30] have evaluated the phenotypic characteristics of tissue fibrocytes *in vivo*. In an animal model of wound healing,[28] numerous CD13$^+$ collagen I$^+$ fibrocytes could be isolated from digested fragments of wounded tissue between 4 and 7 days post-wounding. Collagen I expression was confirmed with actual procollagen I gene expression data. While only a few fibrocytes showed α-SMA immunoreactivity at day 4 post-wounding, on the average, 58.7% of these cells were found to express the myofibroblast marker at day 7 post-wounding. The expression of α-SMA was confirmed at the gene level and the differentiation of fibrocytes into myofibroblast-like cells was associated with a progressive down-regulation of the expression of CD34 and CD45.

In the other study,[30] labeled fibrocytes were tracked in a mouse model of chronic asthma and airway remodeling. In mice systemically sensitized to ovalbumin and chronically exposed to this allergen, labeled fibrocytes accumulated in the airway wall and localized to areas of ongoing collagen deposition below the airway epithelium. The labeled fibrocytes recovered from the airway wall tissue at 24 hours after the last allergen exposure had a phenotype different from the phenotype of the fibrocytes which had been injected intravenously before the allergen exposure. They expressed α-SMA while injected fibrocytes did not, and showed increased collagen I immunoreactivity. Although more than 90% of the injected fibrocytes were CD34-positive cells, only about 40% of the labeled fibrocytes, recovered from the airway wall tissue at 24 hours post-allergen exposure, still expressed the hematopoietic marker. Taken together, the results of these studies provide direct evidence that fibrocytes differentiate into mature fibroblasts and myofibroblasts at the tissue sites under inflammatory conditions and represent an important source of newly produced collagen. They also indicate that fibrocytes rapidly lose the surface markers which can be presently used to differentiate these cells from resident fibroblasts and myofibroblasts. Thus, $CD45^+$ $CD34^+$ collagen I^+ α-SMA^- fibrocytes that have completed their differentiation into fibroblasts and myofibroblasts at the tissue sites likely become $CD45^-$ $CD34^-$ collagen I^+ α-SMA^- and $CD45^-$ $CD34^-$ collagen I^+ α-SMA^+, respectively, and at this stage the origin of these cells is difficult to demonstrate.

Fibrocytes in Asthma

The thickening of the *lamina reticularis* in asthmatic airways occurs at an early stage of the disease and is associated with an increased number of fibroblasts and myofibroblasts in this area.[17,20,40–42] In allergic asthmatics, the number of myofibroblasts further increases in the bronchial mucosa within 24 hours following allergen exposure.[43] These cells have been considered as deriving from pre-existing fibroblasts[43,44] or from airway smooth muscle cells that migrate towards the epithelial basement membrane and de-differentiate to myofibroblasts[43] under the effects of cytokines and growth factors released from epithelial cells and inflammatory cells.

Another explanation of the rapid increase in bronchial myofibroblasts induced by allergen inhalation has been offered by a study[30] demonstrating the appearance of fibrocytes expressing CD34 in conjunction with procollagen I mRNA in the airways of patients with allergic asthma between 4 and 24 hours after the inhalation of the clinically relevant allergen. At 24 hours following allergen inhalation, a substantial proportion of the $CD34^+$ procollagen I $mRNA^+$

cells also expressed α-SMA and localized to areas of collagen deposition below the epithelial basement membrane.[30] It is of note that only a few of the fibroblasts (procollagen I mRNA-positive cells) and myofibroblasts (α-SMA-positive cells, excluding vessels and smooth muscle cells) detected in the airways of the asthmatic patients before allergen inhalation or after the inhalation of the diluent of the allergen alone (control inhalation) expressed the CD34 antigen. By contrast, $CD34^+$ fibrocytes accounted, on the average, for about 40% of the nonvascular, non-smooth muscle cells producing collagen I and differentiating into myofibroblasts in the airways of the same patients at 24 hours following allergen inhalation.

An analysis of the contents of ET-1 and the active form of TGF-β_1 was performed in the bronchoalveolar lavage fluid from patients who underwent an allergen inhalation challenge in the same way as the patients included in the study mentioned above,[30] according to a standard protocol.[45] The results indicate that allergen inhalation induces a progressive increase in the release of these molecules into the airway mucosa between 4 and 24 hours, in comparison with the inhalation of the allergen diluent alone (control inhalation) (Figs. 2A and 2B). Thus, the allergen-induced accumulation of fibrocytes in the bronchial mucosa is paralleled by an increased production of ET-1 and the active form of TGF-β_1 The peak of ET-1 and TGF-β_1 immunoreactivity in the bronchoalveolar lavage fluid is observed at 24 hours following allergen inhalation (Figs. 2A and 2B), at the time when many of the $CD34^+$ procollagen I $mRNA^+$ cells present in the bronchial mucosa following allergen inhalation have been shown to express α-SMA.[30]

In asthmatic airways, the major sources of ET-1 are epithelial cells,[7,46] which also account for the majority of cells producing TGF-β_1[32] together with eosinophils.[33] Because both ET-1 and TGF-β_1 promote the differentiation of human fibrocytes into myofibroblasts *in vitro*,[30] it is possible that the $CD34^+$ procollagen I $mRNA^+$ α-SMA^+ cells present in the airways of asthmatic patients at 24 hours following allergen inhalation[30] are fibrocytes undergoing differentiation into myofibroblasts under the effect of the ET-1 and TGF-β_1 produced in excess by epithelial cells and inflammatory cells.

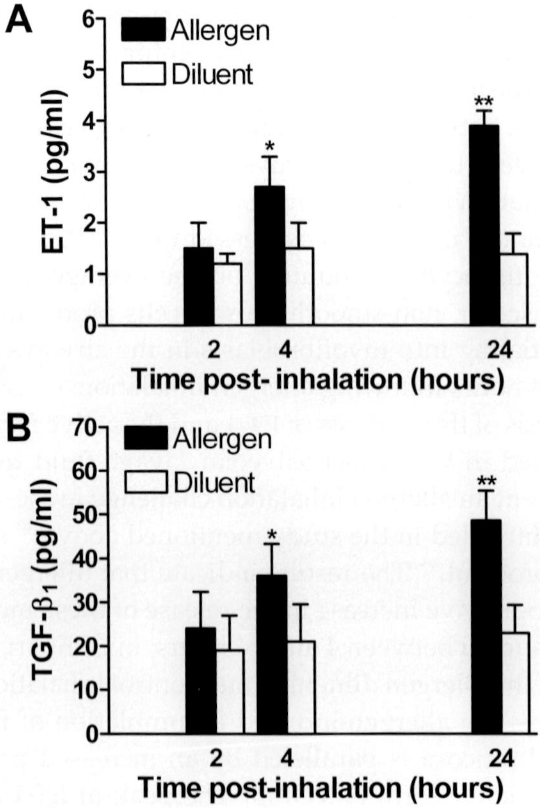

Fig. 2. Kinetics of the release of ET-1 (**A**) and the active form of TGF-β_1 (**B**) in the airways of patients with allergic asthma who underwent two separate inhalation challenges with the allergen to which they were sensitized or with the allergen diluent alone (control inhalation challenge), according to a standard protocol.[30,45] The concentrations of immunoreactive ET-1 and TGF-β_1 were measured by enzyme-linked immunosorbent assay in the bronchoalveolar lavage fluid obtained upon fiberoptic bronchoscopy at each indicated time point after the inhalation of the allergen or the diluent alone. One group of seven patients was tested at each time point. Group data are expressed as the mean + SEM. $*P < 0.05$, $**P < 0.01$ in comparison with control inhalation challenge.

Further information on the role of fibrocytes in asthma has been provided by a recent study,[47] where $CD34^+$ $CD45^+$ α-SMA^+ cells were detected in bronchial biopsies from untreated patients with mild disease. These fibrocytes differentiating into myofibroblasts appeared in clusters close to the epithelial basement membrane and their number correlated with the thickness of the *lamina reticularis*. Moreover, similar cells were isolated from cultures of cells obtained by bronchoalveolar lavage and they showed the migratory capacity and other characteristics of the previously described "activated mobile fibroblasts".[48] The correlation between the number of α-SMA$^+$ fibrocytes and the thickness of the *lamina reticularis* supports the hypothesis that fibrocytes contribute to the pathogenesis of the subepithelial fibrosis in asthma. It is, however, unlikely that conclusive data about this point can be obtained from asthmatic patients because human studies can only provide a snapshot of what is undoubtedly a complex and dynamic process and largely depend on statistical correlations for the interpretation of the data.

While the identification of certain cells in sections of bronchial biopsy specimens from asthmatic individuals provides the only *in vivo* evidence that those cells are involved in the disease process, it should be noted that the procedure carries the risk of false identification of cells co-localized by double labeling. For instance, the conventional microscopy used to identify CD34$^+$ procollagen I mRNA$^+$ cells in one of the study mentioned above[30] may not be able to resolve overlapping or juxtaposed cells and these may be erroneously identified as a single cell with co-localized markers. However, the experiments that demonstrated the accumulation of fibrocytes in the bronchial mucosa of patients with mild asthma[47] were carried out by using confocal fluorescence microscopy, which offers an adequate tool for excluding overlapping cells.

Fibrocytes in Asthma Models

An evaluation of the kinetics of the allergen-induced accumulation of fibrocytes in the airways in relation to the development of subepithelial fibrosis has been performed by using an animal model of allergic asthma that recapitulates most of the inflammatory and structural alterations of the human disease, including the accumulation of eosinophils within and below the airway epithelium and the thickening of the subepithelial zone, with deposition of collagen and other extracellular matrix proteins.[30] In this model, systemically immunized BALB/c mice were challenged with an aerosolized solution of 2.5% ovalbumin (OVA) in phosphate buffered saline (PBS) in a whole body inhalation chamber for 20 minutes, 3 times a week, at intervals of 24 hours, over a period of 8 weeks. Control mice were exposed to the OVA vehicle alone (PBS). During repeated OVA exposures, sensitized mice showed a progressive increase in the number of $CD34^+$ procollagen I^+ cells in the airway wall in comparison with control animals exposed to PBS, as assessed by identification of the double-labeled cells under confocal fluorescence microscopy. In the airway wall of animals exposed to the allergen for 6 to 8 weeks, there were also numerous cells expressing the CD34 antigen in conjunction with α-SMA.

The time-course of fibrocyte accumulation in the airway wall of mice chronically exposed to OVA was examined in relation to the kinetics of two key events: the production of TGF-β_1 by airway resident cells and the deposition of collagen I below the epithelial basement membrane.[30,49] In the epithelium and subepithelial area of mice chronically exposed to OVA, there was a marked increase in the number of cells showing TGF-β_1 immunoreactivity in comparison with mice chronically exposed to PBS for similar periods of time.[49] The peak of TGF-β_1 immunoreactivity was observed between 6 and 8 weeks of chronic exposure to OVA, when many of the $CD34^+$ procollagen I^+ cells also expressed α-SMA,[30,49] suggesting that a substantial proportion of fibrocytes were differentiating into myofibroblasts and that TGF-β_1 was involved in the differentiation process. Interestingly, in mice chronically exposed to OVA both the increase in TGF-β_1 immunoreactivity and the increase in the number of $CD34^+$ procollagen I^+ α-SMA^+ cells occurred in concomitance with the excessive deposition of collagen I below the epithelial basement membrane.[30,49] In view of this correlation, and considering that TGF-β_1 markedly enhances collagen release in cultured $CD34^+$ collagen I^+ fibrocytes acquiring α-SMA expression,[30] it is reasonable to think that the fibrocytes undergoing differentiation into myofibroblasts in the subepithelial area were an important source of that collagen and contributed to the development of subepithelial fibrosis.

Potential Fibrocyte Chemoattractants in Asthma

Human fibrocytes express the receptors for various chemokines, including CXCR4 and CCR7.[26,50] These receptors are also expressed by fibrocytes isolated from BALB/c[26,49] and C57BL/6 mice.[36,37] Evaluation of the ability of CXCL12 (also named stromal cell-derived factor-1α), the ligand for CXCR4, to induce fibrocyte migration *in vitro* and *in vivo* has provided conflicting results.[26,36] One of the putative ligands for CCR7, CCL21 (also named 6Ckine and secondary lymphoid cytokine), has been shown to induce the migration of fibrocytes *in vitro* and *in vivo*.[26] In agreement with these findings, the systemic administration of a neutralizing antibody against CCL21 significantly reduced the accumulation of fibrocytes in the lungs of BALB/c mice in an animal model of allergic asthma.[49] It should be noted, however, that inhibition of CCL21 in this model was also associated with a marked reduction in the accumulation of activated T lymphocytes in the airway wall. Therefore, an indirect effect on the proliferation and differentiation of fibrocytes or their precursors, primarily caused by T lymphocyte depletion, could not be excluded.

The results of another study in mice[37] have suggested that CCR2 may also mediate the recruitment of fibrocytes to the lungs after fibrotic injury, but the relevance of these findings to human fibrotic diseases is unclear given that human fibrocytes isolated from the peripheral blood do not express CCR2,[50,51] unless they are induced to differentiate into cells of a different lineage.[51] In addition, in that study fibrocytes were developed *in vitro* from explants of injured lung tissue and there was no attempt to investigate the accumulation of these cells in the injured lungs *in vivo*.

The *in vivo* evaluation of factors potentially involved in the recruitment of fibrocytes to the tissue site is limited by the expression of chemokine receptors on many resident and inflammatory cells and also by the fact that the number of peripheral blood fibrocytes is very low and insufficient to drive the accumulation of these cells in the lungs after injury. One of the hypotheses proposed to explain this disparity is that the circulating pool of fibrocytes is continuously replenished from the bone marrow.[36] The following observations support this hypothesis: first, following the intravenous injection of labeled fibrocytes, these cells rapidly disappear from the circulation[30] and localize to areas of ongoing collagen deposition at the tissue sites;[26,30,36] secondly, increased numbers of $CXCR4^+$ $CD45^+$ collagen I^+ cells can be detected in the bone marrow of mice during the development of an experimentally induced lung fibrosis associated with the accumulation of $CXCR4^+$ $CD45^+$ collagen I^+ cells in the lungs.[36] Taken together, these findings suggest that fibrocytes can be isolated as such from the bone marrow and can traffic to the lungs after tissue injury. However, the fact that injected fibrocytes can be attracted to the site of tissue injury does not imply that tissue fibrocytes actually derive from a circulating pool of fibrocytes. In addition, in the study demonstrating the presence of $CXCR4^+$ $CD45^+$ collagen I^+ in the bone marrow and in the lungs,[36] these cells were only identified by flow cytometry and there was no attempt to confirm collagen I expression at the gene level.

Another possibility is that fibrocytes represent the intermediate stage of differentiation of a fibroblast/myofibroblast hematopoietic precursor of the monocyte/macrophage lineage at the tissue site. According to this hypothesis, fibrocytes emerge *in vivo* from the same precursor from which they can be obtained *in vitro*[26,27] and it is this precursor, rather than mature fibrocytes, that is predominantly recruited to the tissue sites after injury. Interestingly, both CXCR4- and CCR2-mediated signals in murine bone marrow determine the frequency of $CD34^+$ $CD45^+$ hematopoietic progenitors in the circulation.[52,53] Moreover, the systemic administration of neutralizing antibodies against CXCL12 or CXCR4 may markedly reduce the release of progenitor cells into the circulation[54,55] and mice lacking CCR2 have a paucity of circulating progenitors of the monocyte lineage.[53] Thus, it is possible that the reduction in the accumulation of fibrocytes observed in the murine injured lungs[36] or in cultures on tissue explants from murine injured lungs[37] after inhibition of systemic CXCR4- and CCR2-mediated signals,[36,37] respectively, is due to inhibition of the release of fibrocyte precursors from the bone marrow and not to inhibition of fibrocyte recruitment from the circulation. This hypothesis would also explain why CXCL12 does not induce fibrocyte migration *in vivo*,[26] while the intravenous administration of a neutralizing antibody against CXCL12 is associated with a reduction in the accumulation of fibrocytes in injured lungs.[36]

Some extracellular matrix proteins that are contained in the *lamina reticularis* may be potentially involved in the intraparenchymal migration of fibrocytes in asthmatic airways. Proteolytic fragments of collagen, elastin and fibronectin are known to induce the directed movements of fibroblasts and monocytes/macrophages.[56] Fibronectin is a normal constituent of the *lamina reticularis* and has been shown to attract cultured murine lung fibrocytes in a standard *in vitro* chemotaxis assay,[37] although the intact molecule exhibits a weak chemotactic activity.[56] Fibronectin is extremely susceptible to proteolytic degradation, and some proteolytic fragments produced during tissue repair are those that mediate cell adhesion and migration.[56] It seems likely that these active fragments with high chemotactic activity are also released as a result of the inflammatory process in the airways of asthmatic individuals. Therefore, fibronectin may contribute to the direct movement of fibrocytes towards the area below the epithelial basement membrane, where they have been found to accumulate in asthma.[30,47] Because fibrocytes constitutively produce fibronectin, these cells may then represent an important source of new fibronectin once they have migrated close to the *lamina reticularis*, thereby contributing at the same time to the excessive deposition of extracellular matrix proteins in that area and to the recruitment of other fibrocytes.

Conclusions

The data reviewed in this chapter indicate that bone marrow-derived fibrocytes may contribute to the bronchial (myo)fibroblast population in asthma and may be involved in the genesis of subepithelial fibrosis through the release of excessive amounts of collagen and other extracellular matrix proteins below the bronchial epithelium. Bronchial fibrocytes may derive from a circulating pool of cells that traffic to the lungs as a result of the inflammatory reaction, or may represent the intermediate stage of differentiation of a (myo)fibroblast hematopoietic precursor of the monocyte/macrophage lineage that is recruited to the bronchial mucosa and complete the differentiation into (myo)fibroblast at the tissue site by interaction with activated T lymphocytes and under the influence of fibrogenic cytokines produced by epithelial cells and inflammatory cells, particularly ET-1 and TGF-β_1 (Fig. 3). The factors

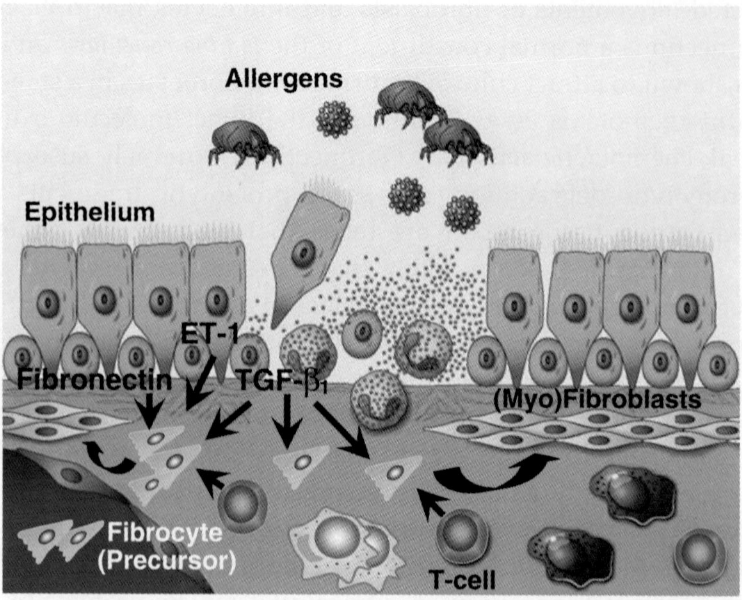

Fig. 3. Proposed role of fibrocytes in the pathogenesis of subepithelial fibrosis in asthma. During the repeated cycles of airway inflammation and repair that occur in asthmatic airways, fibrocytes may be recruited as such from the circulation or may represent the intermediate stage of differentiation of a (myo)fibroblast precursor attracted into the bronchial mucosa as a result of the inflammatory process. The development of fibrocytes from this precursor and/or the differentiation of fibrocytes into fibroblasts/myofibroblasts may occur by interaction with activated T lymphocytes and under the influence of cytokines and growth factory released in excessive amounts by epithelial cells and inflammatory cells, particularly ET-1 and TGF-β_1. Proteolytic fragments of extracellular matrix molecules, such as fibronectin, are generated in the *lamina reticularis* as a result of the inflammatory process and may direct the movement of fibrocytes towards the zone beneath the airway epithelium, where they complete their differentiation into (myo)fibroblasts and produce new extracellular matrix molecules.

involved in the recruitment of fibrocytes or their precursors to the bronchial mucosa in asthma are largely unknown, but it is possible that proteolytic fragments of fibronectin, which are known to be produced during the course of inflammatory processes, mediate in part the migration of fibrocytes to the subepithelial area (Fig. 3).

Further studies investigating how fibrocytes emerge in asthmatic airways and what their fate is may uncover key mechanisms involved in the pathogenesis of airway remodeling and may help clarify the functional role of the remodeling process in asthma.

References

1. Naylor B. (1962) The shedding of the mucosa of the bronchial tree in asthma. *Thorax* **17**: 69–72.
2. Laitinen LA, Heino M, Laitinen A, *et al.* (1985) Damage of the airway epithelium and bronchial reactivity in patients with asthma. *Am Rev Respir Dis* **131**: 599–606.
3. Montefort S, Djukanovic R, Holgate S, Roche WR. (1993) Ciliated cell damage in the bronchial epithelium in asthmatics and non-asthmatics. *Clin Exp Allergy* **23**: 185–189.
4. Shahana S, Bjornsson E, Ludviksdottir D, *et al.* (2005) Ultrastucture of bronchial biopsies from patients with allergic and non-allergic asthma. *Respir Med* **99**: 429–443.
5. Soloperto M, Mattoso VL, Fasoli A, Mattoli S. (1991) A bronchial epithelial cell-derived factor in asthma that promotes eosinophil activation and survival as GM-CSF. *Am J Physiol* **260**: L530–L538.
6. Marini M, Vittori E, Hollemborg J, Mattoli S. (1992) Expression of the potent inflammatory cytokines, granulocyte-macrophage colony-stimulating factor and interleukin-6 and interleukin-8, in bronchial epithelial cells of patients with asthma. *J Allergy Clin Immunol* **89**: 1001–1009.
7. Vittori E, Marini M, Fasoli A, *et al.* (1992) Increased expression of endothelin in bronchial epithelial cells of asthmatic patients and effects of corticosteroids. *Am Rev Respir Dis* **146**: 1320–1325.
8. Mori L, Kleimberg J, Mancini C, *et al.* (1995) Bronchial epithelial cells of atopic patients with asthma lack the ability to inactivate allergens. *Biochem Biophys Res Commun* **217**: 817–824.
9. Mattoli S. (2001) Allergen-induced generation of mediators in the mucosa. *Environ Health Perspect* **109**(Suppl 4): 553–557.

10. Bucchieri F, Puddicombe SM, Lordan JL, *et al.* (2002) Asthmatic bronchial epithelium is more susceptible to oxidant-induced apoptosis. *Am J Respir Cell Mol Biol* **27**: 179–185.
11. Bayram H, Rusznak C, Khair OA, *et al.* (2002) Effect of ozone and nitrogen dioxide on the permeability of bronchial epithelial cells in cultures of non-asthmatic and asthmatic subjects. *Clin Exp Allergy* **32**: 1285–1292.
12. Wark PA, Johnston SL, Bucchieri F, *et al.* (2005) Asthmatic bronchial epithelial cells have a deficient innate immune response to infection with rhinovirus. *J Exp Med* **201**: 937–947.
13. Kay AB. (1991) Asthma and inflammation. *J Allergy Clin Immunol* **87**: 893–910.
14. Bousquet J, Chanez P, Lacoste JY, *et al.* (1992) Asthma: A disease remodeling the airways. *Allergy* **47**: 3–11.
15. Elias JA, Zhu Z, Chupp G, Homer J. (1999) Airway remodeling in asthma. *J Clin Invest* **104**: 1001–1006.
16. Fish JE, Peters SP. (1999) Airway remodeling and persistent airway obstruction in asthma. *J Allergy Clin Immunol* **104**: 509–516.
17. Gabbrielli S, Di Lollo S, Stanflin N, Romagnoli P. (1994) Myofibroblasts and elastic and collagen fiber hyperplasia in the bronchial mucosa: a possible basis for the progressive irreversibility of airflow obstruction in asthma. *Pathologica* **86**: 157–160.
18. Stewart AG, Tomlinson PR, Wilson J. (1993) Airway wall remodeling in asthma: a novel target for the development of anti-asthma drugs. *Trend Pharmacol Sci* **14**: 275–279.
19. Sobonya RE. (1984) Quantitative structural alterations in long-standing allergic asthma. *Am Rev Respir Dis* **130**: 289–292.
20. Brewster CEP, Howarth PH, Djukanovic R, *et al.* (1990) Myofibroblasts and subepithelial fibrosis in bronchial asthma. *Am J Respir Cell Mol Biol* **3**: 507–511.
21. Hoshino M, Nakamura Y, Sim JJ. (1998) Expression of growth factors and remodeling of the airway wall in asthma. *Thorax* **53**: 21–27.
22. Cokugras H, Akcakaya N, Seckin S, *et al.* (2001) Ultrastructural examination of bronchial biopsy specimens from children with moderate asthma. *Thorax* **56**: 25–29.
23. Roche WR, Beasley R, Williams JH, Holgate ST. (1989) Subepithelial fibrosis in the bronchi of asthmatics. *Lancet* **1**(8637): 520–524.
24. Gabbiani G. (2003) The myofibroblast in wound healing and fibrocontractive diseases. *J Pathol* **200**: 500–503.

25. Bucala R, Spiegel LA, Chesney J, *et al.* (1994) Circulating fibrocytes define a new leukocyte subpopulation that mediates tissue repair. *Mol Med* **1**: 71–81.
26. Abe R, Donnelly SC, Peng T, *et al.* (2001) Peripheral blood fibrocytes: differentiation pathway and migration to wound sites. *J Immunol* **166**: 7556–7562.
27. Yang L, Scott PG, Giuffre J, *et al.* (2002) Peripheral blood fibrocytes from burn patients: Identification and quantification of fibrocytes in adherent cells cultured from peripheral blood mononuclear cells. *Lab Invest* **82**: 1183–1192.
28. Mori L, Bellini A, Stacey MA, *et al.* (2005) Fibrocytes contribute to the myofibroblast population in wounded skin and originate from the bone marrow. *Exp Cell Res* **304**: 81–90.
29. Frid MG, Brunetti JA, Burke DL, *et al.* (2006) Hypoxia-induced pulmonary vascular remodeling requires recruitment of circulating mesenchymal precursors of a monocyte/macrophage lineage. *Am J Pathol* **168**: 659–669.
30. Schmidt M, Sun G, Stacey MA, *et al.* (2003) Identification of circulating fibrocytes as precursors of bronchial myofibroblasts in asthma. *J Immunol* **171**: 380–389.
31. Redington A, Madden J, Frew A, *et al.* (1997) Transforming growth factor $\beta 1$ in asthma: measurement in bronchoalveolar lavage fluid. *Am J Respir Crit Care Med* **156**: 642–647.
32. Vignola AM, Chanez P, Chiappara G, *et al.* (1997) Transforming growth factor-beta expression in mucosal biopsies in asthma and chronic bronchitis. *Am J Respir Crit Care Med* **156**: 591–599.
33. Minshall EM, Leung DY, Martin RJ, *et al.* (1997) Eosinophil-associated TGF-beta1 mRNA expression and airway fibrosis in bronchial asthma. *Am J Respir Cell Mol Biol* **17**: 326–333.
34. Mattoli S, Soloperto M, Marini M, Fasoli A. (1991) Levels of endothelin in the bronchoalveolar lavage fluid of patients with symptomatic asthma and reversible airflow obstruction. *J Allergy Clin Immunol* **88**: 376–384.
35. Hay DWP, Henry PJ, Goldie RG. (1996) Is endothelin-1 a mediator in asthma? *Am J Respir Crit Care Med* **154**: 1594–1597.
36. Phillips RJ, Burdick MD, Hong K, *et al.* (2004) Circulating fibrocytes traffic to the lungs in response to CXCL 12 and mediate fibrosis. *J Clin Invest* **114**: 438–446.

37. Moore BB, Kolodsick JE, Thannickal VJ, *et al.* (2005) CCR2-mediated recruitment of fibrocytes to the alveolar space after fibrotic injury. *Am J Pathol* **166**: 675–684.
38. Dubé J, Chakir J, Dubé C, *et al.* (2000) Synergistic action of endothelin-1 on the activation of bronchial fibroblasts isolated from normal and asthmatic subjects. *Int J Exp Path* **81**: 429–437.
39. Glassberg MK, Ergul A, Wanner A, Puett D. (1994) Endothelin-1 promotes mitogenesis in airway smooth muscle cells. *Am J Respir Cell Mol Biol* **10**: 316–321.
40. Ward C, Pais M, Bish R, *et al.* (2002) Airway inflammation, basement membrane thickening and bronchial hyperresponsiveness in asthma. *Thorax* **57**: 309–316.
41. Jeffery PK. (2001) Remodeling in asthma and chronic obstructive lung disease. *Am J Respir Crit Care Med* **164**: S28–S38.
42. Payne DN, Rogers AV, Adelroth E, *et al.* (2003) Early thickening of the reticular basement membrane in children with difficult asthma. *Am J Respir Crit Care Med* **167**: 78–82.
43. Gizycki MJ, Adelroth E, Rogers AV, *et al.* (1997) Myofibroblast involvement in the allergen-induced late response in mild atopic asthma. *Am J Respir Cell Mol Biol* **16**: 664–673.
44. Davies DE, Wicks J, Powell RM, *et al.* (2003) Airway remodeling in asthma: New insights. *J Allergy Clin Immunol* **111**: 215–225.
45. Brown JR, Kleimberg J, Marini M, *et al.* (1998) Kinetics of eotaxin expression and its relationship to eosinophil accumulation and activation in bronchial biopsies and bronchoalveolar lavage (BAL) of asthmatic patients after allergen inhalation. *Clin Exp Immunol* **114**: 137–146.
46. Ackerman V, Carpi S, Bellini A, *et al.* (1995) Constitutive expression of endothelin in bronchial epithelial cells of patients with symptomatic and asymptomatic asthma and modulation by histamine and interleukin-1. *J Allergy Clin Immunol* **96**: 618–627.
47. Nilsson K, Larsen K, Hultgard-Nilsson A, *et al.* (2005) Localization of fibrocytes in bronchial biopsies from mild untreated asthmatic patients. *Proc Am Thorac Soc* **2**(Abstracts Issue): A513.
48. Larsen K, Nilsson K, Tufvesson E, *et al.* (2005) Presence of activated mobile fibroblasts in bronchial alveolar lavage from mild asthmatic patients. *Wound Rep Reg* **13**(1): A24.
49. Mattoli S. (2006) Tissue repair in asthma: the origin of airway subepithelial fibroblasts and myofibroblasts. In C Chaponnier, A

Desmoulière & G Gabbiani (eds), *Tissue Repair, Contraction and the Myofibroblast*, pp. 40–46. Landes Bioscience and Springer Science + Business Media, Georgetown.

50. Pilling D, Buckley CD, Salmon M, Gomer RH. (2003) Inhibition of fibrocyte differentiation by serum amyloid P. *J Immunol* **171**: 5537–5546.
51. Hong KM, Burdick RJ, Heber PD, Strieter RM. (2005) Characterization of human fibrocytes as circulating adipocyte progenitors and the formation of human adipose tissue in SCID mice. *FASEB J*. doi: 10.1096/fj.05-4295fje.
52. Lapidot T, Petit I. (2002) Current understanding of stem cell mobilization: the role of chemokines, proteolytic enzymes, adhesion molecules, cytokines, and stromal cells. *Exp Hematol* **30**: 973–981.
53. Serbina NV, Pamer EG. (2006) Monocyte emigration from bone marrow during bacterial infection requires signals mediated by chemokine receptor CCR2. *Nat Immunol* **7**: 311–317.
54. Petit I, Szyper-Kravitz M, Nagler A, *et al.* (2002) G-CSF induces stem cell mobilization by decreasing bone marrow SDF-1 and up-regulating CXCR4. *Nat Immunol* **3:** 687–694.
55. Sweeney EA, Lortat-Jacob H, Priestley GV, Nakamoto B, Papayannopoulou T. (2002) Sulfated polysaccharides increase plasma levels of SDF-1 in monkeys and mice: involvement in mobilization of stem/progenitor cells. *Blood* **99**: 44–51.
56. Briggs SL. (2005) The role of fibronectin in fibroblast migration during tissue repair. *J Wound Care* **14**: 284–287.

Chapter 7

Fibrocytes and Other Fibroblast/Myofibroblast Progenitors in Systemic Sclerosis

Arnold E. Postlethwaite*

Systemic sclerosis (SSc), also called scleroderma, is a fibrosing autoimmune disease. There are several different clinical types of SSc that have variable internal organ fibrosis and degrees of skin involvement with fibrosis. A prominent vasculopathy often precedes the onset of fibrosis. The origins of myofibroblasts which are the dominant matrix synthesizing cell in SSc involved tissue has traditionally been thought to arise from transformation of resident fibroblasts left over from embryonic development in response to cytokine/growth factor stimulation from activated T cells and monocytes that populate areas of fibrosis in SSc organs. We have evidence that the blood mononuclear cells generate large numbers of fibroblast-like cells when they are cultured with an auto antigen for SSc, type I collagen. Development of myofibroblasts from epithelial mesenchymal

*Goodman Chair of Excellence Professor of Medicine, Division of Connective Tissue Diseases, University of Tennessee Health Science Center, 956 Court Avenue, Room G326, and Veterans Administration Medical Center, Memphis, TN 38163, USA. Tel.: (901) 448-5774; Fax: (901) 449-7265. E-mail: apostlethwai@utmem.edu

transition (EMT), pericytes and fibrocytes are discussed as possible contributors to fibrosis in SSc. Implication of alternative sources of myofibroblasts in SSc from the traditionally held view of their development from resident fibroblasts for therapeutic approaches is discussed.

Systemic Sclerosis Clinical Characteristics

Systemic sclerosis (SSc) is an autoimmune disease the consequence of which is multiorgan fibrosis. Three major clinical forms of SSc are recognized based on the pattern of dermal fibrosis.[1] "Diffuse SSc" is characterized by dermal fibrosis of the proximal and distal extremities, face, back, chest and trunk. Fibrosis of the lungs, esophagus, stomach, small and large intestines, and heart are common. "Limited SSc" is characterized by dermal fibrosis of the distal upper (below elbows) and distal lower (below knees) portions of the extremities, face and upper anterior chest. There is less internal organ fibrosis but pulmonary hypertension and primary biliary cirrhosis not infrequently occur late in the disease. The term "scleroderma" is often used interchangeably with SSc, but the former term denotes only skin fibrosis and not the true systemic fibrotic nature of the condition. A less common form of SSc is SSc sine scleroderma in which there is internal organ fibrosis but no dermal fibrosis.

The Vasculature in SSc

Vascular changes in SSc include vasomotor instability, small vessel structural abnormalities and intravascular abnormalities.[2] Raynaud's phenomenon, or vasospastic episodes of digital arteries, is extremely common in SSc and can result in digital infarction.[1] Raynaud's phenomenon usually predates the onset of clinically detectable fibrosis by several months, suggesting vascular abnormalities are an early event in the evolution of SSc. Vasospasm has also been documented in arteries supplying the internal organs.[3]

There is evidence of microvascular injury with elevated levels of circulating vascular endothelial growth factor (VEGF), soluble vascular adhesion molecule (sVCAM), soluble E-selectin and endothelin 1, and platelet thrombi in the arterioles and venules in patients with SSc.[2,4] Endothelial cells early and throughout the course of SSc exhibit subtle changes in the plasma membrane, roughening of the surface, vacuolization, swelling, necrosis, sloughing of cells and reduplication of the basement membrane.[2]

The Immune System in SSc

The autoimmune nature of SSc is evident from the presence of anti-nuclear, anti-nucleolar, anti-topoisomerase I (Scl-70) and anti-centromere antibodies, and T cell immunity to a variety of body components, including types I, III and IV collagens, RNA, proteoglycan, mitochondia, etc. (see Tables 1 and 2).[5–7]

Table 1 Autoantigens to which Antibodies are Raised in SSc

• DNA Topisomerase I	• IL-6
• Cytomeric protein	• IL-8
• PM-Scl antigen	• IgG
• Myenteric neurons	• Cardiolipin
• RNA polymerases I, II and III	• Fc γR
• Endothelial cells	• Histones
• Fibroblasts	• Mitochondria
• Smooth muscle	• Laminin
• Granulocytes	• Single stranded RNA
• Erythrocytes	• U3 and U11 nuclear ribonucleoproteins
• Platelets	• Th ribonucleoproteins
• Thyroid tissue	• Upstream binding factors (NOR-90)
• Salivary gland tissue	• High motility group (HMG) 17 nucleosome in protein
• Neutrophil cytoplasmic antigens	• Ku antigen
• Heat shock protein	• Fibrillin 1
• Types I, III and IV collagens	

Table 2 Antigens to which there is Cellular Autoimmunity in SSc

- Collagen types I, II, III and IV
- Laminin
- kDRNA polypeptide
- Elastin
- Cell associated antigen
  - — High molecular weight RNA
  - — Muscle cells
  - — Fibroblasts
  - — Epithelial cells
  - — Lymphocyte lysates
  - — Liver microsomer and mitochrondrin
  - — Human myelin basic protein
  - — Thryoglobulin

The Fibroblast Phenotype in SSc

The fibroblast populations identified *in situ* or grown from explants of clinically involved fibrotic skin from patients with SSc are predominantly myofibroblasts.[8] In contrast, fibroblast populations in clinically uninvolved skin *in situ* or grown from explants of clinically uninvolved skin in SSc are composed of nearly normal percentages of myofibroblasts.[8] The fibroblasts from the involved skin of patients with SSc synthesize increased amounts of types I, III, VI and VII collagens, proteoglycans and tissue inhibitor of metalloproteinase (TIMP)-1, but reduced amounts of matrix metalloproteinase (MMP)-1 (type I collagenase).[8–14] There are alterations in several signal transduction pathways in fibroblasts from clinically involved cultured skin, including elevated TGF-β receptor types I and III, Smad 2/3, Smad 7, αvβ 5 integrin, endothelin receptor A and B, angiotensin II receptor type I, PDGF receptor α and β, necdin, phokinase, C delta, EKG/MAPG and in mammalian target of rapamycin pathway (mTOR).[15]

Fibroblasts cultured from the involved skin of patients with SSc have increased levels of intracellular interleukin (IL-1α), and when stimulated with IL-1, tumor necrosis factor (TNF-α) or basic fibroblast growth factor (bFGF) overexpress intracellular interleukin-1 receptor antagonist (icIL-1ra) protein.[16] The overexpression of icIL-1ra in normal fibroblasts produces a myofibroblast phenotype and suppresses MMP-1 production in response to stimulation with TNF-α and may be important in the development and maintenance of the myofibroblast phenotype in fibrotic SSc skin.[17]

Accumulation of T cells, Monocytes and Mast Cells in Clinically Involved Skin in SSc

Biopsies of clinically involved skin in SSc often show accumulation in perivascular and adjacent tissue locations of $CD4^+$ T cells, $CD14^+$ monocytes and mast cells.[18,19] The numbers of $CD14^+$ monocytes in the dermis directly correlate with the degree of fibrosis.[20] These cells are capable of synthesis of a variety of fibrogenic cytokines, including IL-4, TGF-β and mast cell tryptase that stimulate fibroblast proliferation, chemotaxis and synthesis of collagen, proteoglycan and fibronectin.[5,21] The T cells in such dermis have the hallmark of a clonally expanded population suggesting that they are reacting to an antigen present in the involved tissue.[22]

Relationship of Autoimmunity, Vascular Abnormalities and Fibrosis in SSc (the Old Paradigm)

There are no complete animal models of SSc that reproduce the three major components of the disease, i.e. vascular abnormalities, autoimmunity and multiorgan fibrosis, and it has been difficult to explore pathogenetic mechanisms operative in SSc. Therefore, we are left with results of studies conducted *in vitro* with specimens obtained from patients with the disease to form the basis for developing paradigms to explain the pathogenesis of SSc. The most commonly accepted paradigm by researches in the SSc field is that endothelial damage by unknown events leads to tissue accumulations of T cells, monocytes and mast cells, which in turn elaborate cytokines that expand the resident fibroblast population, cause myofibroblast development (via TGF-β and other cytokines) and upregulate the synthesis of collagen, proteoglycans, TIMP, and other matrix components and suppress MMP-1 production.[5,21]

There also may be a role for certain autoantibodies such as anti-PDGF R which stimulate Ha-Ras-ERK112 and ROS cascades and type I collagen-gene expression and myofibroblast conversion of normal human fibroblasts (Ref. 23; and anti-fibroblast antibody which stimulates fibroblasts to upregulate expression of ICAM-1, IL-6, IL-1α and IL-1β[24]).

Possible Alternative Sources of Fibroblasts in SSc

The old paradigm relies on resident tissue fibroblasts left over from embryonic development as to the source of excessive matrix deposited in involved organs in SSc. It should be realized that these normal resident fibroblasts are heterogenous such that fibroblasts from different anatomic sites respond differently to various stimuli with regards to the genes they express[25–28] and should be considered distinct cell types.[25] There is evidence from the study of HOX genes that this may in part be due to imprinting.[25] The HOX genes are highly conserved and function to determine positional orientation of differentiating cells during embryogenesis. The HOX genes are similar in fibroblasts from a given location in the adult as in the same anatomical location in the embryo.[25] *In vitro,* myofibroblasts which express α-smooth muscle actin (SMA), the ED-A splice variant of fibronectin (ED-A FN) and Thy-1, can be generated by prolonged exposure of fibroblasts to some cytokines, including TGF-β, by plating at low density or by overexpressing intracellular IL-1 receptor antagonist (IL-1ra).[17,29,30]

Resident Fibroblast Progenitors

A potential source of fibroblasts in SSc is the resident progenitor cells. It has been convincingly demonstrated in several renal fibrosis models in mice that the tubular epithelial cells can transition into fibroblasts.[31] During embryogenesis, epithelia transition into mesenchymal cells by a process called epithelial-mesenchymal transition.[32] During embryogenesis, EMT is an important mechanism by which epithelial cells are released from the surrounding cells and tissue; the cells migrate to the various parts of the embryo to undergo differentiation into other cell types.[32] *In vitro*, EMT has been shown to be initiated by TGF-β, bFGF, epidermal growth factor, and insulin-like growth factor II.[33–36] The renal fibrosis in murine models of SLE is partially mediated by EMT and can be inhibited by treatment of the mice or *in vitro* active by bone morphogenic protein-7.[37] In SSc lung fibrosis, alveolar epithelial cells could by EMT contribute fibroblasts to the pool being stimulated to synthesize excessive matrix. In the gastrointestinal tract in SSc, gut epithelial cells might be a source of fibroblasts participating in fibrogenesis of the gastrointestinal track.

Hepatic stellate cells and their relatives, pericytes, which comprise microvessels, have been shown to transdifferentiate into myofibroblasts.[38–41] Perhaps they contribute to portal cirrhosis seen in late limited SSc. Indeed, in SSc, there is evidence from carefully performed immunohistochemical analyses of skin biopsies that both the pericytes and fibroblast-like cells expressing ED-A FN, Thy1 and α-SMA are present in the involved skin but not in the uninvolved skin from patients with SSc.[39]

Fibroblast Progenitors from the Circulation in Patients with SSc and Related Fibrotic Conditions

At the beginning of this chapter, the salient features of SSc were reviewed to give the reader an understanding of the nature of the vascular damage, and inflammatory and immune reactions and compartmentalization of fibrosis to regions of the body surface and selected internal organs in the major subtypes of the disease, i.e., limited and diffuse SSc. Just why certain areas of the skin are involved in limited versus diffuse SSc, why there is variation in the degree of skin fibrosis in the involved skin of patients with SSc and why only certain organs are predisposed to fibrosis has been a mystery. Circulating fibroblast progenitors could home to sites where endothelial damage has occurred, and via transvessel migration, accumulate in the perivascular spaces. As discussed below, activated T cells and monocyte could provide cytokine stimulation that would effect transdifferentiation of the fibroblast precursors into fibroblasts and myofibroblasts.

Circulating Fibrocytes and other Progenitors of Fibroblast-like Cells (FLC)

Fibrocytes have been the most studied circulating precursor to FLC as has been reviewed elsewhere in this chapter. There are also other progenitors of spindle-shaped cells or FLC that have been described. Labat and coworkers reported that cultured blood monocytes from patients with osteomyelosclerosis and Engelmann disease are spontaneously transformed into FLC which he termed "neofibroblasts".[42] Subsequent studies revealed that these neofibroblasts were also derived from HLA-DR^+ monocytes.[43] These neofibroblasts could differentiate into other cell types and were found to secrete type I collagen, uromodulin, amyloid-β peptide, α-fetoprotein and carcinoembryonic antigen.[44] Neofibroblasts could be generated from normal donors by exposure of monocytes to soluble factors from T cells[42,43] in culture for only 17 days and would revert back to a macrophage phenotype when in contact with T cells.[42,43]

Kuwana and co-workers described a FLC which he termed "monocyte-derived mesenchymal progenitors" (MOMPs)[45] that were derived from $CD14^+$ monocytes from normal donors.[45] MOMPs, like fibrocytes, express CD45, CD34 and type I collagen, but unlike fibrocytes, express CD14.[45] MOMPs were differentiated from human PBMC cultures only when the cultures were preformed on fibronectin-coated surfaces with low glucose Dulbecco's modified Eagle's medium (DMEM) supplemented with 10% fetal calf serum and required exposure to soluble factors from the $CD14^-$ PBMC population.[45] MOMPs could be differentiated into other mesenchymal cell types such as osteoblasts, adipocytes, myocytes and chondrocytes, when exposed to appropriate growth factors *in vitro*.[45] MOMPs also expressed, in addition to the above surface markers,

CD13, CD11b, CD11c, CD64, HLA Class I and HLA-DR, CD40, CD86, CD29, CD44, CD54, CD105/SH2, CD31, CD144, Flt-1, Ac-LDL, type III collagen, fibronectin and vimentin.[45]

Zhao and co-workers characterized another FLC called "pluripotent stem cell" or PSC.[46] PSC were grown from $CD14^+$ enriched human mononuclear cells in eight well Lab Tech Chamber slides coated with collagen and repeatedly stimulated with monocyte-colony stimulating factor (MCSF) and leukemia inhibitory factor (LIF), and assumed a spindle-shaped morphology after seven days in culture. PSCS stained positive for CD14, CD35, CD45 and type I collagen.[46] Although further description was given as to cell surface markers, PSCs were shown to differentiate into $CD3^+/CD8^+$ T cells, epithelial cells, neuronal cells and hepatocytes when cultured with appropriate growth and differentiating factors.[46] Exposure to lipsopolysacharide converted PSCs back to macrophages.[46]

It is not readily apparent from the literature, how and if fibrocytes, neofibroblasts, MOMPs and PSCs are related. Clearly MOMPs and PSCs express CD14, a property not shared with fibrocytes.[47] It is intriguing that fibrocytes can differentiate into adiopocytes.[48]

Circulating fibroblast/myofibroblast progenitors might be contributing to the pathogenesis of SSc and could explain distinct patterns and degrees of fibrosis in limited versus diffuse SSc. The heterogeneity of the vascular endothelium with organ specific characteristics might also play a role in determining whether endothelial cells in a particular organ are more resistant to injury in the initial and chronic phases of SSc.[49] The endothelium of some organs might be susceptible to a given injury and resistant to other types of injury. The endothelium, vasculature of skin and other organs when injured in the SSc patient would permit circulating $CD4^+$ T cells, $CD14^+$ monocytes and fibrocytes to adhere to the damaged endothelium and the transvessel migrate into the perivascular tissue of the organ where transdifferentiation into the matrix synthesizing myofibroblasts could occur. $CD4^+$ T cells $CD14^+$ monocytes could effect cytokine production after activation by autoantigens such as type I collagen prevalent in all tissue and organs involved in fibrosis in SSc.

In an interesting article by Pilling and co-workers, it was reported that serum amyloid protein inhibits the outgrowth of fibrocytes from PBMC and that sera from patients with SSc had reduced levels of the amyloid protein.[50] We have been assessing the outgrowth of FLC from the PBMC population of patients with SSc with and without stimulation with the auto-antigen type I collagen. We have found that the PMBC from patients with SSc generate large numbers of FLC which are derived from the $CD14^+$ monocyte population activated by soluble factors from type I collagen-stimulated PBMC.[51] We are in the process of characterizing these FLC, and they are different from PSCs, MOMPs and fibrocytes in that they are $CD14^+/CD34^-$/type I $collagen^+$ and do not revert to monocytes upon exposure to lipopolysaccharide and are not pluirpotent (unpublished data). Studies are in progress to characterize these progenitors of FLC from CI-stimulated SSc PBMC.

Overall Hypothetical Scheme for Pathogenesis of SSc

In Fig. 1, a theoretical scheme for SSc immunopathogenesis is illustrated. The scheme attempts to summarize events likely to occur in the disease that results in fibrosis in specific organs. The earliest events in the pathogenesis of SSc likely involve some sort of injury to the endothelium of small blood vessels (viral, bacterial, toxic, chemical). An appropriate genetic background is likely essential for the vascular injury to lead to SSc disease. Vasomotor instability might be the result of the injury and may contribute to the injury via release of endothelium 1 from the endothelium.

Once the endothelium is injured and activated, adhesion molecules may anchor circulating lymphocytes, monocytes and mast cells. These cells will then migrate through the damaged vessels wall and accumulate in the perivascular tissue. Monocytes/dendritic cells/macrophages interacting with antigens in the tissue such as ubiquitous type I collagen present antigens to T cells leading to T cell autoreactivity and generation of auto-antibodies from T cell dependent B cell activation.

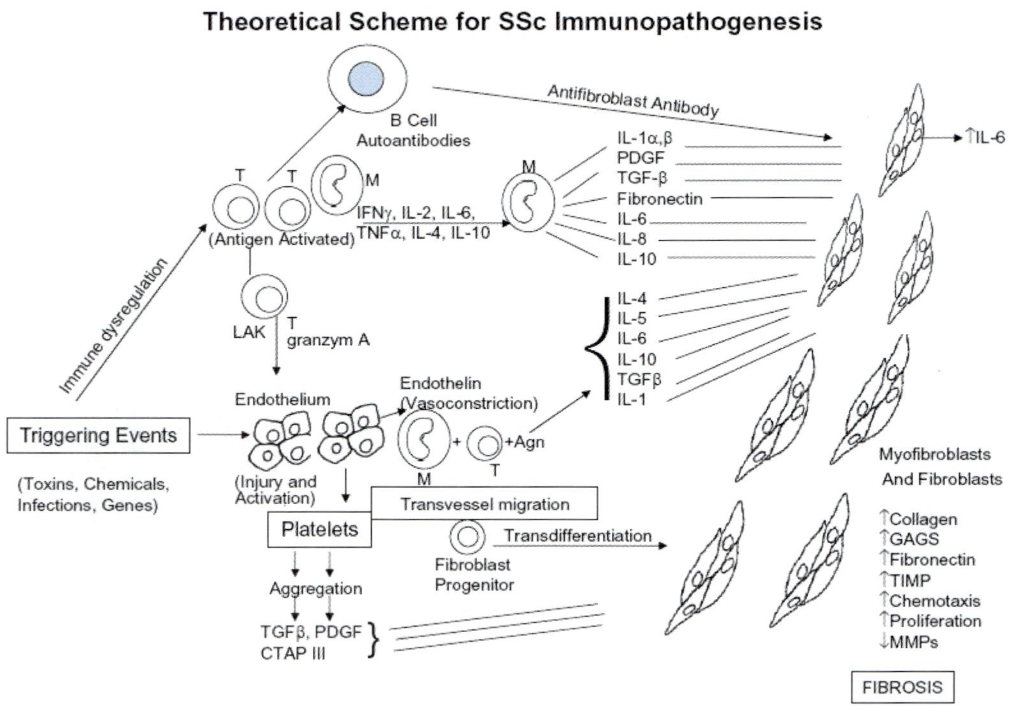

Fig. 1. Events trigger endothelial injury that leads to the transvessel migration of T cells, B cells, monocytes and mast cells. Platelets are activated to release fibrogenic and inflammatory mediators. T cells and monocytes/dendritic cells interact with autoantigens or other external antigens resulting in cytokine production and generation of auto-antibody producing B cells and plasma cells. Anti-fibroblast antibody and anti-PDGF R antibody can activate matrix production by fibroblasts. Activated T cells and monocytes produce fibrogenic cytokines that regulate the fibroblast/myofibroblast phenotype and increase matrix production. Circulating progenitors for fibroblasts/myofibroblasts migrate through damaged vessels and undergo transdifferentiation to matrix producing fibroblasts/myofibroblasts.

Not shown in Fig. 1 is the generation of fibroblasts/myofibroblasts by EMT of the epithelial cells in organs such as the lungs, GI tract and kidneys via exposure of epithelial cells to TGF-β and other cytokines/growth factors such as insulin-like growth factor 1 and bFGF.

Platelets, by interacting with exposed type I and III collagens in injured vessels and via other activators, release TGF-β, PDGF and other growth factors and mediators to facilitate accumulation of monocytes and promote EMT and myofibroblast development.

The induction of anti-reactive T cells and auto-antibody producing B cells and plasma cells potentiates fibrosis by release of cytokines from activated T cells that activate monocytes to release TGF-β and other cytokines important in EMT; generation of myofibroblasts; and transdifferentiation of $CD14^+$ monocytes and fibrocytes into fibroblasts/myofibroblasts.

Anti-fibroblast and anti-PDGF-R antibodies produced by B cells could activate fibroblasts to synthesize increased amounts of collagen and other matrix components.

Types I, III, IV collagens generated by the expanded population of fibroblasts and myofibroblasts, would produce not only tissues fibrosis and failure of organ function which is the hallmark of SSc, but also more auto-antigens that would be presented to $CD4^+$ T cells by monocytes/dendritic cells. Recurring injury to the small vessel endothelium would insure a renewable supply of fibroblasts/myofibroblasts from circulating precursors to perpetuate the fibrosis of SSc.

New Treatment Strategies for SSc based on Circulating Fibroblast Progenitors

Clearly more work needs to be done to establish a pathogenetic role of circulating fibroblast precursors in SSc. The *in vitro* culture techniques to generate FLC from blood progenitor cells lends itself to screening of potential compounds to inhibit the process at various stages. Identification of cytokines and other agents that might inhibit EMT or FLC transdifferentiation from monocyte or fibrocyte progenitors is a necessary step in developing treatments that could eventually be taken to the clinic and tried in patients with SSc.

References

1. Seibold JR. (1993) *Connective Tissue Diseases Characterized by Fibrosis.* W.B. Saunders, Philadelphia, pp. 1113–1143.
2. Kahaleh MB, LeRoy EC. (1998) *Vascular Factors in the Pathogenesis of Systemic Sclerosis.* John Wiley and Sons Ltd., New York, NY, pp. 107–118.
3. Furst DE, Davis JA, Clements PJ, *et al.* (1981) Abnormalities of pulmonary vascular dynamics and inflammation in early progressive systemic sclerosis. *Arthritis Rheum* **24**(11): 1403–1408.
4. Kuryliszyn-Moskal A, Klimiuk PA, Sierakowski S. (2005) Soluble adhesion molecules (sVCAM-1, sE-selectin), vascular endothelial growth factor (VEGF) and endothelin-1 in patients with systemic sclerosis: relationship to organ systemic involvement. *Clin Rheumatol* **24**(2): 111–116.
5. Postlethwaite AE. (1995) Role of T cells and cytokines in effecting fibrosis. *Int Rev Immunol* **12**(2–4): 247–258.
6. Bernstein RM, Steigerwald JC, Tan EM. (1982) Association of antinuclear and antinucleolar antibodies in progressive systemic sclerosis. *Clin Exp Immunol* **48**(1): 43–51.
7. Schur PH, Monroe M, Rothfield N. (1972) The gammaG subclass of antinuclear and antinucleic acid antibodies. *Arthritis Rheum* **15**(2): 174–182.
8. Kirk TZ, Mark ME, Chua CC, *et al.* (1995) Myofibroblasts from scleroderma skin synthesize elevated levels of collagen and tissue inhibitor of metalloproteinase (TIMP-1) with two forms of TIMP-1. *J Biol Chem* **270**(7): 3423–3428.
9. Gay RE, Buckingham RB, Prince RK, *et al.* (1980) Collagen types synthesized in dermal fibroblast cultures from patients with early progressive systemic sclerosis. *Arthritis Rheum* **23**(2): 190–196.
10. Bashey RI, Perlish S, Nochumson S, *et al.* (1977) Connective tissue synthesis by cultured scleroderma fibroblasts. II. Incorporation of 3h-glucosamine and synthesis of glycosaminoglycans. *Arthritis Rheum* **20**(3): 879–885.

11. Higuchi T, Ohnishi K, Hayashi H, *et al.* (1994) Changes in skin disaccharide components correlate with the severity of sclerotic skin in systemic sclerosis. *Acta Derm Venereol* **74**(3): 179–182.
12. Takeda K, Hatamochi A, Ueki H, Nakata M, Oishi Y. (1994) Decreased collagenase expression in cultured systemic sclerosis fibroblasts. *J Invest Dermatol* **103**(3): 359–363.
13. Rudnicka L, Varga J, Christiano AM, *et al.* (1994) Elevated expression of type VII collagen in the skin of patients with systemic sclerosis. Regulation by transforming growth factor-beta. *J Clin Invest* **93**(4): 1709–1715.
14. Peltonen J, Kahari L, Uitto J, Jimenez SA. (1990) Increased expression of type VI collagen genes in systemic sclerosis. *Arthritis Rheum* **33**(12): 1829–1835.
15. Pannu J, Trojanowska M. (2004) Recent advances in fibroblast signaling and biology in scleroderma. *Curr Opin Rheumatol* **16**(6): 739–745.
16. Higgins GC, Wu Y, Postlethwaite AE. (1999) Intracellular IL-1 receptor antagonist is elevated in human dermal fibroblasts that overexpress intracellular precursor IL-1 alpha. *J Immunol* **163**(7): 3969–3975.
17. Kanangat S, Postlethwaite AE, Higgins GC, Hasty KA. (2006) Novel functions of intracellular IL-1ra in human dermal fibroblasts: Implications in the pathogenesis of fibrosis. *J Invest Dermatol* **126**(4): 756–765.
18. Fleischmajer R, Perlish JS, Reeves JR. (1977) Cellular infiltrates in scleroderma skin. *Arthritis Rheum* **20**(4): 975–984.
19. Roumm AD, Whiteside TL, Medsger TA, Jr., Rodnan GP. (1984) Lymphocytes in the skin of patients with progressive systemic sclerosis. Quantification, subtyping, and clinical correlations. *Arthritis Rheum* **27**(6): 645–653.
20. Kraling BM, Maul GG, Jimenez SA. (1995) Mononuclear cellular infiltrates in clinically involved skin from patients with systemic sclerosis of recent onset predominantly consist of monocytes/macrophages. *Pathobiology* **63**(1): 48–56.
21. White B. (1996) Immunopathogenesis of systemic sclerosis. *Rheum Dis Clin North Am* **22**(4): 695–708.
22. Sakkas LI, Xu B, Artlett CM, *et al.* (2002) Oligoclonal T cell expansion in the skin of patients with systemic sclerosis. *J Immunol* **168**(7): 3649–3659.
23. Baroni SS, Santillo M, Bevilacqua F, *et al.* (2006) Stimulatory autoantibodies to the PDGF receptor in systemic sclerosis. *N Engl J Med* **354**(25): 2667–2676.

24. Chizzolini C, Raschi E, Rezzonico R, *et al.* (2002) Autoantibodies to fibroblasts induce a proadhesive and proinflammatory fibroblast phenotype in patients with systemic sclerosis. *Arthritis Rheum* **46**(6): 1602–1613.
25. Chang HY, Chi JT, Dudoit S, *et al.* (2002) Diversity, topographic differentiation, and positional memory in human fibroblasts. *Proc Natl Acad Sci USA* **99**(20): 12877–12882.
26. Muller GA, Rodemann HP. (1991) Characterization of human renal fibroblasts in health and disease: I. Immunophenotyping of cultured tubular epithelial cells and fibroblasts derived from kidneys with histologically proven interstitial fibrosis. *Am J Kidney Dis* **17**(6): 680–683.
27. Garrett DM, Conrad GW. (1979) Fibroblast-like cells from embryonic chick cornea, heart, and skin are antigenically distinct. *Dev Biol* **70**(1): 50–70.
28. Dugina V, Alexandrova A, Chaponnier C, *et al.* (1998) Rat fibroblasts cultured from various organs exhibit differences in alpha-smooth muscle actin expression, cytoskeletal pattern, and adhesive structure organization. *Exp Cell Res* **238**(2): 481–490.
29. Serini G, Gabbiani G. (1999) Mechanisms of myofibroblast activity and phenotypic modulation. *Exp Cell Res* **250**(2): 273–283.
30. Masur SK, Dewal HS, Dinh TT, *et al.* (1996) Myofibroblasts differentiate from fibroblasts when plated at low density. *Proc Natl Acad Sci USA* **93**(9): 4219–4223.
31. Kalluri R, Neilson EG. (2003) Epithelial-mesenchymal transition and its implications for fibrosis. *J Clin Invest* **112**(12): 1776–1784.
32. Hay ED. (1995) An overview of epithelio-mesenchymal transformation. *Acta Anat* (Basel) **154**(1): 8–20.
33. Fan JM, Ng YY, Hill PA, *et al.* (1999) Transforming growth factor-beta regulates tubular epithelial-myofibroblast transdifferentiation *in vitro*. *Kidney Int* **56**(4): 1455–1467.
34. Okada H, Danoff TM, Kalluri R, Neilson EG. (1997) Early role of Fsp1 in epithelial-mesenchymal transformation. *Am J Physiol* **273**(4 Pt 2): F563–F574.
35. Morali OG, Delmas V, Moore R, *et al.* (2001) IGF-II induces rapid beta-catenin relocation to the nucleus during epithelium to mesenchyme transition. *Oncogene* **20**(36): 4942–4950.
36. Strutz F, Zeisberg M, Ziyadeh FN, *et al.* (2002) Role of basic fibroblast growth factor-2 in epithelial-mesenchymal transformation. *Kidney Int* **61**(5): 1714–1728.

37. Zeisberg M, Bottiglio C, Kumar N, *et al.* (2003) Bone morphogenic protein-7 inhibits progression of chronic renal fibrosis associated with two genetic mouse models. *Am J Physiol Renal Physiol* **285**(6): F1060–F1067.
38. Cassiman D, Libbrecht L, Desmet V, *et al.* (2002) Hepatic stellate cell/myofibroblast subpopulations in fibrotic human and rat livers. *J Hepatol* **36**(2): 200–209.
39. Rajkumar VS, Howell K, Csiszar K, *et al.* (2005) Shared expression of phenotypic markers in systemic sclerosis indicates a convergence of pericytes and fibroblasts to a myofibroblast lineage in fibrosis. *Arthritis Res Ther* 7(5): R11130–R1123.
40. Ivarsson M, Sundberg C, Farrokhnia N, *et al.* (1996) Recruitment of type I collagen producing cells from the microvasculature *in vitro*. *Exp Cell Res* **229**(2): 336–349.
41. Sundberg C, Ivarsson M, Gerdin B, Rubin K. (1996) Pericytes as collagen-producing cells in excessive dermal scarring. *Lab Invest* **74**(2): 452–466.
42. Labat ML, Bringuier AF, Seebold C, *et al.* (1991) Monocytic origin of fibroblasts: spontaneous transformation of blood monocytes into neo-fibroblastic structures in osteomyelosclerosis and Engelmann's disease. *Biomed Pharmacother* **45**(7): 289–299.
43. Labat ML, Bringuier AF, Seebold-Choqueux C, *et al.* (1991) Cystic fibrosis: production of high levels of uromodulin-like protein by HLA-DR blood monocytes differentiating towards a fibroblastic phenotype. *Biomed Pharmacother* **45**(9): 387–401.
44. Bringuier AF, Seebold-Choqueux C, Moricard Y, *et al.* (1992) T-lymphocyte control of HLA-DR blood monocyte differentiation into neo-fibroblasts. Further evidence of pluripotential secreting functions of HLA-DR monocytes, involving not only collagen but also uromodulin, amyloid-beta peptide, alpha-fetoprotein and carcinoembryonic antigen. *Biomed Pharmacother* **46**(2-3): 91–108.
45. Kuwana M, Okazaki Y, Kodama H, *et al.* (2003) Human circulating $CD14^+$ monocytes as a source of progenitors that exhibit mesenchymal cell differentiation. *J Leukoc Biol* **74**(5): 833–845.
46. Zhao Y, Glesne D, Huberman E. (2003) A human peripheral blood monocyte-derived subset acts as pluripotent stem cells. *Proc Natl Acad Sci USA* **100**(5): 2426–2431.

47. Bucala R, Spiegel LA, Chesney J, *et al.* (1994) Circulating fibrocytes define a new leukocyte subpopulation that mediates tissue repair. *Mol Med* **1**(1): 71–81.
48. Hong KM, Burdick MD, Phillips RJ, *et al.* (2005) Characterization of human fibrocytes as circulating adipocyte progenitors and the formation of human adipose tissue in SCID mice. *Faseb J* **19**(14): 2029–2031.
49. Aird WC. (2003) Endothelial cell heterogeneity. *Crit Care Med* **31**(4 Suppl): S221–S230.
50. Pilling D, Buckley CD, Salmon M, Gomer RH. (2003) Inhibition of fibrocyte differentiation by serum amyloid P. *J Immunol* **171**(10): 5537–5546.
51. Postlethwaite AE, Wong WK, Ingels J, *et al.* (2004) Increased outgrowth of fibroblast-like cells (FLC) from SSc collagen stimulated peripheral blood mononuclear cells (PBMC) is associated with decreased lung carbon monoxide diffusion capacity (DLCO). In: *Arthritis and Rheumatism Annual Meeting* 2004, San Antonio, TX, p. S426.

Chapter 8

Fibrocytes in Interstitial Lung Disease

Brigitte N. Gomperts[†] and Robert M. Strieter[*]

It is now over a decade since Bucala and colleagues first described a population of cells in circulation with fibroblast-like properties that were involved in tissue repair. Since that time we have learned a significant amount about these bone marrow-derived cells that are now termed "fibrocytes," and which contribute to wound healing and fibrosis. These cells express leukocyte markers such as CD34, CD45 and CD13, and have also been demonstrated to express mesenchymal markers such as procollagens I and III, vimentin and fibronectin. In addition, they have also been shown to express the chemokine receptors, CXCR4 and CCR7 that appear to be important in cellular trafficking from the vascular to the extravascular compartment. Fibrocytes have been shown to contribute to a number of fibrotic

[*]Address for Correspondence: Robert M. Strieter, M.D. Tel.: (310)-794-1999; Fax: (310)-794-1998. E-mail: rstrieter@mednet.ucla.edu

Department of Internal Medicine, University of Virginal School of Medicine, P.O. Box 800466, Charlottesvilla, VA 22908-0466, USA. Tel.: (434)-982-3297; Fax (434)-979-4967. E-mail: rms4w@virginia.edu

[†]Department of Pediatrics, University of California, Los Angeles, School of Medicine, Los Angeles, CA 90045.

disorders, and here we review their involvement in lung diseases including pulmonary fibrosis, asthma and vascular remodeling.

Introduction

The Fibrocyte is a Unique Cell Population that has been Implicated in Wound Repair

While a number of cells have been implicated in tissue injury and repair, the fibroblast/myofibroblast plays a pivotal role for the generation of the extracellular matrix (ECM), which serves as the foundation for re-establishment of tissue integrity. Bucala and associates in 1994 discovered a unique fibroblast-like cell.[1] The cells were blood-borne, not a leukocyte, and their presence in wounds was not due to local infiltration of surrounding connective tissue fibroblasts.[1] This discovery led to the identification of a novel and distinct population of blood-borne cells with fibroblast-like features that were subsequently named "fibrocytes".[1] By scanning electron microscopy, fibrocytes are morphologically distinct from leukocytes and display prominent cell surface projections, that are intermediate in size between microvilli and pseudopodia.[1] Abe and colleagues were able to demonstrate that peripheral blood fibrocytes could be cultured from a $CD14^+$ cell population.[2] However, it remains unclear whether these cells are derived from $CD14^+$ cells, as differentiated fibrocytes from peripheral blood are characterized as spindle-shaped, negative for CD14 and nonspecific esterase stain (i.e. not monocytes or macrophages); and negative for cell surface markers for epithelial and endothelial cells.[1,3] Fibrocytes comprise only 0.1% to 0.5% of the nonerythrocyte cells in peripheral blood when assessed for dual collagen and CD45 or CD34 expression by FACS analysis.[3–5] Fibrocytes in culture (i.e. presence of serum) spontaneously begin to express alpha-smooth muscle actin (α-SMA), and will further markedly express levels of α-SMA in the presence of either TGF-β or endothelin, compatible with differentiation into myofibroblasts.[2–6] Fibrocytes stain positive for the fibroblast markers, vimentin; collagens I and III, and fibronectin.[1,3,5] Fibrocytes are negative for CD3, CD4, CD8, CD16, CD19, CD25 and CD54, but express the adhesion molecules, CD11b and CD18.[1,2,4,5] In addition, fibrocytes stain positive for the common leukocyte antigen (CD45RO), the pan-myeloid antigen (CD13), HLA-DR, and the hematopoietic stem cell antigen, CD34.[1–6]

While early time-points of fibrocyte culture are associated with the expression of $CD34^+$, $CD45^+$, Col I^+, and $vimentin^+$, differentiation to myofibroblast-like cells after exposure to TGF-β or endothelin results in a concomitant gain in expression of α-SMA and loss of expression of CD34[3,5–7] and CD45.[3] These cellular changes support the notion that with differentiation, these cells lose their "stem" and common "leukocyte" markers. Therefore, the hallmark markers for these cells in circulation are $CD45^+$, $CD34^+$, Col I^+, and $vimentin^+$.[1,3,5] Although fibrocytes, when initially isolated, are $CD34^+$, and CD34 expression was originally identified as a cell marker of hematopoietic stem cells, it remains to be fully elucidated whether fibrocytes are actually derived from hematopoietic or mesenchymal progenitor or stem cells.[1,4,5]

Fibrocytes have been found to be pleiotropic in their behavior, and their function in generating constituents of extracellular matrix makes them apropos in promoting fibrosis. They are potent antigen presenting cells and can elicit the recruitment and activation of T cells.[8] Fibrocytes can participate in promoting angiogenesis by inducing an angiogenic phenotype in cultured endothelial cells and orchestrate angiogenesis *in vivo*.[9] Fibrocytes can secrete chemokines, cytokines, and growth factors that are relevant in mediating fibroproliferation.[4,10] Fibrocytes have been found localized in various tissues under both normal and pathologic conditions.[4]

Fibrocyte Trafficking

Trafficking of leukocytes involves a complicated array of adhesion molecules, chemoattractants and chemoattractant receptors to allow the leukocyte to exit the bone marrow and be recruited to a specific tissue region.[11] Classic cell trafficking has been well described for leukocytes, but is an area of relatively new investigation for fibrocytes. It is likely that there is significant overlap between the molecules and signaling pathways of leukocytes and fibrocytes.

The chemokine biological system is made up of chemoattractant ligands and their seven transmembrane receptors, and chemokines are by far the most diverse group of cytokines involved in leukocyte/cellular trafficking.[11] They trigger leukocyte adhesion to endothelial cells and are also involved in leukocyte migration. The diversity of chemokine receptor expression on subpopulations of leukocytes, as well as the specific temporal and spatial patterns of expression of chemokine ligands provides a mechanism for directing cellular immune responses.[11] The complicated, multi-step process of trafficking of a leukocyte from the bone marrow into tissues at a distant site therefore involves specific combinations of chemokine ligands and chemokine receptors to orchestrate all these events. The lung has characteristic expression of chemokine ligands at defined points after injury to mediate the recruitment of specific cells at specific time points.

Human fibrocytes have been shown to express the chemokine receptors CCR3, CCR5, CCR7 and CXCR4[2,5] (Table 1) (gene chip and superarray data of fibrocyte chemokine gene expression — unpublished data). In contrast, mouse fibrocytes have been shown to express CCR7, CXCR4 and CCR2.[2,5,12] CXCR4 has been shown to be the most pivotal chemokine receptor in hematopoietic and non-hematopoietic stem cell homing, and the differential expression of CXCL12 in different tissues creates a gradient that is essential for

Table 1 Chemokine Receptors and Ligands in Trafficking of Fibrocytes

CXC RECEPTOR	**LIGANDS**
CXCR4	CXCL12
CC RECEPTORS	**LIGANDS**
CCR2	CCL2, CCL12
CCR3	CCL5, CCL7, CCL8, CCL11, CCL13, CCL15, CCL24, CCL26, CCL28
CCR5	CCL3, CCL3L1, CCL4, CCL5, CCL8, CCL14
CCR7	CCL19, CCL21

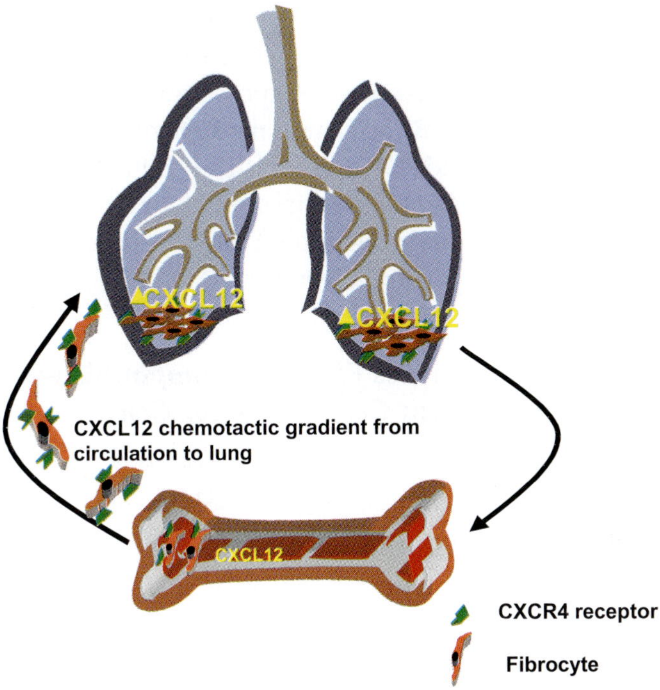

Fig. 1. Fibrocytes in pulmonary fibrosis.

trafficking of $CXCR4^+$ cells (e.g. leukocytes).[13] Fibrocytes have been shown to express CXCR4, consistent with their progenitor/stem cell phenotype and to migrate in response to CXCL12 under specific conditions *in vitro*.[3] Moreover, the biological axis of CXCL12/CXCR4 has been demonstrated to play a major role in mediating the extravasation and contribution of fibrocytes to pulmonary fibrosis[3] (Fig. 1). Fibrocytes also express CCR7, which is a chemokine receptor that is important in dendritic cell and T cell migration in, for example, draining lymph nodes that express the chemokine ligands for the CCR7 receptor, CCL21 and CCL19. Phillips and colleagues identified a population of fibrocytes that expressed CCR7 in bleomycin-induced pulmonary fibrosis, which were distinct from the CXCR4 expressing fibrocytes.[3] They noted that the intrapulmonary recruitment of $CD45^+Col\ I^+CXCR4^+$ fibrocytes was greater than that of $CD45^+Col\ I^+CCR7^+$ fibrocytes, which correlated with collagen deposition in the lungs of bleomycin-exposed mice.[3] In mice, CXCR4, CCR7, and potentially CCR2 appear to mediate recruitment of fibrocytes to the lung.[3,12] If indeed these cells can traffic to human lung, become activated, proliferate and differentiate into myofibroblasts, then preventing their initial recruitment into the lung would impact on the pathogenesis of pulmonary fibrosis.

The Fibrocyte Demonstrates Plasticity Compatible with the Concept of an Adult Stem Cell/Progenitor Cell

An important definition as to whether a cell exhibits adult stem cell/progenitor cell properties is the ability of the cell to demonstrate plasticity. Fibrocytes have been shown to transition to α-SMA^+ cell in the presence of serum or increased doses of TGF-β.[2–6] In addition, fibrocytes can behave in a similar manner as preadipocytes from visceral or subcutaneous adipose tissue, as fibrocytes can differentiate into adipocytes [14]. Interestingly, specific ECM molecules (i.e. type I collagen) and TGF-β can inhibit fibrocyte to adipocyte differentiation via down-regulation of PPARγ.[14] During fibrocyte to adipocyte differentiation, cDNA microarray analysis revealed gene expression clusters that were similar for either visceral or subcutaneous preadipocyte-to-adipocyte differentiation.[14] Moreover, fibrocytes that are undergoing adipogenesis can be engrafted into SCID mice and form human adipose tissue.[14] While mouse fibrocytes may express CCR2,[12] Hong and colleagues found very little, to no, CCR2 gene expression on human fibrocytes by microarray analysis and quantitative real-time PCR, and no protein expression of CCR2 by FACS analysis on undifferentiated human fibrocytes.[14] However, CCR2 gene expression and CCR2 protein expression was markedly induced and expressed when human fibrocytes were exposed to adipogenic differentiation media/conditions.[14] The addition of a PPARγ antagonist to the adipogenic media/conditions significantly attenuated the increase in CCR2 expression, which suggested that PPARγ was important for the expression of CCR2 in these cells.[14]

Furthermore, the CCR2 receptor expression was found to be functional, as fibrocytes expressing CCR2 treated with adipogenic media had a significant migratory response to the CCR2 ligand, CCL2.[14]

Thus the paradigm of plasticity supports the notion that fibrocytes can be transformed into other mesenchymal lineage cells compatible with the concept that these cells are circulating mesenchymal progenitor cells (CMPC). Therefore in the context of lung fibroproliferative disorders, developing strategies to attenuate the recruitment of CMPC into lung tissue will impact on whether these cells integrate into the lung and contribute to pulmonary fibrosis.

Fibrocytes in Pulmonary Fibrosis

Pulmonary Fibrosis

Idiopathic interstitial pneumonias (IIPs) are a heterogeneous group of diffuse parenchymal lung disorders resulting from injury to the lung parenchyma and associated with varying degrees of inflammation and fibrosis.[15,16] IIPs can be classified into seven distinct entities based on clinical manifestations, pathology, and radiologic features.[16] One of these entities, idiopathic pulmonary fibrosis (IPF), is a progressive and fatal disease. IPF is defined as a chronic fibrosing form of IIP limited to the lungs and associated with biopsy proven pathology showing a histologic pattern of usual interstitial pneumonia (UIP).[15,16] UIP is not unique to IPF, and has been reported to be associated with asbestosis, chronic hypersensitivity pneumonitis, and collagen vascular disorders with associated interstitial lung disease.[17]

While ongoing research continues to investigate multiple hypotheses for the pathogenesis of UIP, neither the natural history nor the exact pathogenesis of UIP is currently known. Whether the pathology of UIP begins as UIP and remains unchanged with disease progression or begins as a cellular infiltrative pattern that progresses over time to UIP remains unclear. UIP is diagnosed on the basis of temporally heterogeneous areas of normal lung, active fibrosis, and end-stage honeycomb fibrosis.[15] This histopathology suggests differences in kinetics of pathological events that are occurring at different points in time that ultimately lead to dysregulated repair associated with aberrant vascular remodeling and marked deposition of ECM leading to fibrosis. Studies have demonstrated that IIPs do not occur in isolation as separate entities, but they may occur as a continuum with overlapping features of chronic inflammation that leads to end-stage fibrosis. For example, Flaherty and associates demonstrated a significant histopathological variability in surgical lung biopsies from patients with IIP.[18] They found interlobar and intralobar histopathologic variability of IIP with components of chronic inflammation (i.e. NSIP) with more fibrosis (i.e. UIP). Their findings have been further substantiated by Katzenstein and colleagues,[19] who found in a majority of explanted lung specimens concomitant UIP with NSIP. The presence of UIP or discordant UIP with NSIP on biopsy appears to dictate a worse prognosis, as compared to concordant NSIP.[20] Nevertheless, what this notion supports is that evolving pulmonary fibrosis equates to less response to conventional immunosuppressive agents and worse prognosis. Therefore, understanding the host factors that contribute to the continuum that ultimately leads to pulmonary fibrosis associated with IIPs will lead to novel therapies to treat and prevent the development of pulmonary fibrosis. Since pulmonary fibrosis appears to be a dynamic process, the opportunity to target cells that promote fibrosis (i.e. fibroblasts/myofibroblasts) would be a novel strategy to attenuate the development of pulmonary fibrosis.

The Origin of the Fibroblast/Myofibroblast: A Pivotal Cell in Mediating Fibroproliferation in Pulmonary Fibrosis

Currently, one classical and two contemporary concepts exist that fit the origin of fibroblasts/myofibroblasts in lung tissue during the pathogenesis of pulmonary fibrosis.[3,4,21–24] First, the classical concept is that tissue injury induces activation of a resident interstitial fibroblast to differentiate into a myofibroblast that migrates into the intraalveolar space, proliferates, and expresses constituents of the ECM leading to intraalveolar and interstitial pulmonary fibrosis.[22–24] Second, lung injury and changes in the microenvironment of the epithelium including the basement membrane, can induce epithelial cells to transition to a mesenchymal phenotype to become fibroblasts/myofibroblasts, and these cells subsequently contribute to fibroproliferation.[1,2,21,24] Third, circulating fibrocytes (CMPC), that may be derived from bone marrow precursor cells, home and extravasate under specific guidance by chemoattractants into sites of tissue injury, differentiate to myofibroblasts, proliferate, and contribute to the generation of ECM relevant to pulmonary fibrosis.[1,2,4,24]

Phillips and colleagues identified a population of $CD45^+$Col I^+CXCR4^+ human fibrocytes and observed significant chemotaxis of these cells *in vitro* in response to the CXCR4 ligand, CXCL12.[3] They used a murine model of bleomycin-induced pulmonary fibrosis to examine the kinetics and magnitude of migration of these fibrocytes *in vivo*.[3] Purified human fibrocytes were tail-vein injected into SCID mice that had already been exposed to either bleomycin or saline for 4 days.[3] After a further 4 days the mice were sacrificed and the lungs analyzed for the presence of infiltrating human fibrocytes.[3] Significantly greater numbers of human $CD45^+$Col I^+CXCR4^+ fibrocytes were observed in bleomycin-treated lungs in SCID mice, compared to those lungs that received saline alone.[3] Using the bleomycin mouse model in immunocompetent C57Bl/6 mice, they also found that transcription of both procollagen III and procollagen I was dramatically upregulated in mice exposed to bleomycin (up to 30-fold higher) as compared with mice treated with saline alone, and they found an increase in total collagen protein by Sircol assay.[3]

Once they had characterized the kinetics of collagen deposition in C57Bl/6 mice in response to bleomycin, they performed a kinetic analysis of fibrocyte infiltration in the lungs of bleomycin exposed mice.[3] They showed that $CD45^+$Col I^+CXCR4^+ cells began to appear in the lung 2 days after bleomycin treatment, became maximal at day 8, and remained elevated at days 16 and 20.[3] Expression of $CD45^+$Col I^+CXCR4^+ cells in the lung of saline-treated mice also increased initially before returning to the levels observed in naive mice by days 16 and 20.[3] The most likely explanation for the latter phenomenon was that intratracheal instillation of saline itself can promote an inflammatory response. In contrast to the time-dependent accumulation of murine fibrocytes in the lungs of bleomycin-exposed animals, the steady-state levels of circulating $CD45^+Col\ I^+CXCR4^+$ fibrocytes in the blood remained similar in both groups, but were found to be markedly higher in bleomycin exposed animals at days 1 to 4 (unpublished observation). On the basis of finding an extremely early time point of mobilization of fibrocytes in the circulation of bleomycin exposed animals, they next examined whether the early mobilization was due to changes in the potential of bone marrow-derived cells compatible with the fibrocyte phenotype. The number of $CD45^+Col\ I^+CXCR4^+$ fibrocytes found in the bone marrow of bleomycin-challenged animals was markedly greater than the cells found in normal saline control mice.[3] Recently, we have found that fibrocytes can be mobilized into the circulation when mice are exposed to GCSF, MCSF and GMCSF (unpublished data). These data suggest that the bone marrow is at least one potential source of $CD45^+Col\ I^+CXCR4^+$ circulating fibrocytes.

Phillips and colleagues also identified a second fibrocyte population that is $CD45^+Col\ I^+CCR7^+$.[3] These $CCR7^+$ fibrocytes also traffic to the lungs of bleomycin-treated mice; however, the absolute number of $CCR7^+$ fibrocytes found in the fibrotic lung was two-fold to three-fold lower than the number of $CXCR4^+$ fibrocytes present under similar conditions.[3] Thus, these data suggest that although two chemotactic chemokine receptors are present on the surface of fibrocytes, CXCR4 appears predominant for the recruitment of fibrocytes to fibrotic lungs. In addition, the expression of CXCR4 and CCR7 also suggests the potential for two distinct populations of fibrocytes that may have different roles in wound healing.

The CXCR4 ligand, CXCL12, was significantly higher in the lungs of animals exposed to bleomycin for 8 days than the comparable saline control or the naive control.[3] Thus, these data support the notion that a CXCL12 gradient existed between the lungs and plasma of bleomycin-treated mice, which could promote the recruitment of $CD45^+Col\ I^+CXCR4^+$ fibrocytes to the fibrotic lung (Fig. 1). A similar chemokine gradient between the lungs and plasma of bleomycin-treated mice existed for CCL21 (6Ckine), but not for CCL19 (ELC), the putative ligands for CCR7.[3]

Given the importance of the CXCR4/CXCL12 biological axis in fibrocyte recruitment, a strategy to deplete CXCL12 was employed in the bleomycin-lung mouse model. Pulmonary collagen deposition was significantly reduced under conditions of CXCL12 depletion in bleomycin-exposed mice, in comparison with those mice exposed to bleomycin and treated with control antibodies.[3] The addition of neutralizing CXCL12 antibodies to bleomycin-treated lungs did not, however, completely attenuate collagen deposition to the level of the saline control.[3] Taken together, these data indicate that inhibition of the CXCR4/CXCL12 chemotactic axis reduces intrapulmonary recruitment of $CD45^+$Col $I^+$$CXCR4^+$ fibrocytes and significantly abrogates lung fibrosis in bleomycin-exposed mice.

To exclude the possibility that the effect of CXCL12 depletion influenced the infiltration of conventional leukocyte populations, FACS analysis of CD4 and CD8 T cells, NK cells, neutrophils, and monocytes/macrophages was performed from single cell suspensions of bleomycin-exposed lungs in animals treated with anti-CXCL12 or control antibodies.[3] There was no statistical difference in the numbers of CD4 and CD8 T cells, NK cells, neutrophils, and monocytes/macrophages trafficking to the lungs of bleomycin-exposed mice treated with either anti-CXCL12 or control antibodies. Although CXCL12 clearly does mediate recruitment of $CXCR4^+$ leukocytes to the lungs, the fact that neutralizing anti-CXCL12 antibodies did not block migration of leukocytes here can be explained by the observation that these peripheral blood mononuclear cells express chemokine receptors other than CXCR4. It is possible, therefore, that inhibition of a single chemokine receptor/ligand combination (i.e. CXCR4/CXCL12) does not prevent these conventional leukocytes from intrapulmonary infiltration in response to chemokines other than CXCL12. Furthermore, these findings support the notion that blocking the CXCR4/CXCL12 biological axis under conditions of bleomycin-induced pulmonary fibrosis effected only the infiltration of fibrocytes into the lung.

To further verify that systemic addition of anti-CXCL12 antibodies selectively reduced intrapulmonary infiltration of

$CD45^+Col\ I^+CXCR4^+$ fibrocytes and attenuated bleomycin-induced pulmonary fibrosis, both H&E staining and morphometric analysis with a collagen-specific dye (picosirus red) were performed, and the experiment showed that depletion of CXCL12 significantly reduced bleomycin-induced pulmonary fibrosis.[3] In addition, blocking the recruitment of $CD45^+Col\ I^+CXCR4^+$ fibrocytes to bleomycin-exposed lungs, also reduced expression of immunohistochemical expression of α-SMA.[3] This suggests that $CD45^+Col\ I^+CXCR4^+$ fibrocytes recruited from the peripheral circulation may ultimately develop an α-SMA$^+$ phenotype in fibrotic lungs, which is compatible with the *in vitro* findings for human fibrocytes.

Hashimoto and colleagues performed bone marrow transplantation in mice and in so doing transferred GFP labeled bone marrow-derived progenitor cells into wild type mice.[25] They had durable engraftment with >92% of $CD45^+$ cells in the bone marrow being GFP positive. No significant radiation induced pneumonitis was seen. These mice were then exposed to endotracheal bleomycin-injury and analysis of the lungs from these mice showed large clusters of bone marrow-derived GFP positive cells in areas of active fibrosis related to bleomycin exposure. More than 80% of the collagen I expressing cells in the bleomycin-injured lungs were GFP^+ and therefore bone marrow-derived cells. However, dual staining of GFP positive cells in the lung showed almost no staining for α-SMA, implying that in this model and at the time point assessed, the lung myofibroblasts did not appear to arise from bone marrow-derived cells.[25] These findings were in direct contrast to the findings of Philips *et al.*[3] and another GFP bone marrow chimeric mouse model in BALB/c mice, where analysis of GFP positive bone marrow-derived fibrocytes in skin wounds demonstrated clear evidence of the expression of α-SMA in GFP^+ fibrocytes at day 4 post-wounding.[26] As further confirmation that fibrocytes are bone marrow-derived and can differentiate into α-SMA cells, we have transplanted lethally irradiated wild-type mice with GFP^+ bone marrow and subsequently demonstrated that in bleomycin exposed lungs under conditions of CXCL12 depletion, there was a marked reduction in GFP^+ fibrocytes in the lungs of these animals (unpublished data).

Bone marrow-derived progenitor myofibroblasts have also been found in pulmonary organizing alveolitis/fibrosis after lung irradiation in mice.[27] For these studies GFP male bone marrow was transplanted into wild-type female mice and the mice were then exposed to total lung irradiation. Less than 1% of the endothelial cells in the areas of fibrosis were GFP positive and therefore were derived from the bone marrow, but a significant number of GFP positive cells were seen in the fibrotic areas and these cells were also vimentin positive.[27] Bromodeoxyuridine labeling of developing fibrotic areas showed that dividing cells were predominantly GFP^+, Y-chromosome$^+$ and vimentin$^+$. These data suggest that CMPC from the bone marrow are involved in the fibrotic response of the lung to radiation injury and that once recruited these CMPC proliferate at sites of injury and fibrosis. Mice that were subsequently treated with manganese superoxide dismutase-plasmid/liposome intratracheal injection 24 hours before total lung irradiation demonstrated a significant decrease in GFP^+, vimentin$^+$ cells in the injured lung.[27]

Moore and colleagues have examined the contribution of fibrocytes to fibrosis in a FITC-induced lung injury model.[12] In this study fibrocytes were obtained by bronchoalveolar lavage (BAL) from a population of cells that had transmigrated to the airway lumen. In addition, they assessed fibrocytes from minced lung specimens. The BAL cells or minced lungs were cultured and analyzed for the dual expression of collagen I and CD45. Continued culture of these fibrocytes resulted in loss of CD45 expression but a persistence of collagen I expression. They found that populations of fibrocytes from both the B6/129F2 and C57Bl/6 strains of mice expressed CXCR4, CCR5, CCR7 and CCR2.[12] This is in contrast to human fibrocytes which do not seem to express CCR2 after isolation.[14] Fibrocytes isolated from the mouse lungs expressed CCR2, migrated to the CCL2 and CCL12 ligands and lost expression of CCR2 when cultured *in vitro* to a differentiated fibroblast.[12]

Moore and associates[12] treated wild-type and $CCR2^{-/-}$ mice on both the C57BL/6 or B6/129F2 backgrounds with intratracheal FITC and found that the absolute number of lung fibrocytes present in cultures from the BAL of wild-type mice was significantly greater than noted in cultures from $CCR2^{-/-}$ mice.[12] The absolute number of lung fibrocytes in C57BL/6 mice was higher than in B6/129F2 mice, which correlated with the previously observed differences in the magnitude of the fibrotic response to FITC between these two mouse strains. They explained the differences in fibrocyte accumulation in the BAL from wild-type as compared to $CCR2^{-/-}$ mice treated with FITC as reflecting differences in recruitment, as there was no difference in proliferation of these cells in culture.[12] However, CCR2 is highly expressed on macrophage populations, however, no studies were performed to determine whether there were differences in the macrophage or other leukocyte populations in the lungs of wild-type as compared to $CCR2^{-/-}$ FITC-treated mice, which could have contributed to the degree of pulmonary fibrosis in these mice after FITC exposure. In previous studies, Okuma and colleagues have shown that CCR2 knockout mice had less pulmonary fibrosis than wild-type mice after bleomycin exposure, and that this directly correlated with a significant reduction in macrophages in the BAL fluid together with reduced levels of MMP-2 and MMP-9.[28]

Moore and colleagues[12] then performed bone marrow transplants with wild-type bone marrow in lethally irradiated $CCR2^{-/-}$ mice. They showed that recruitment of lung fibrocytes in response to FITC injury was restored in the mice that received $CCR2^{+/+}$ bone marrow. Lung fibrocytes cultured from the $CCR2^{+/+}$ bone marrow transplant mice were positive for CCR2, indicating that they were of donor origin.[12] Conversely, when $CCR2^{-/-}$ mice received a $CCR2^{-/-}$ bone marrow transplant, the mice were protected from FITC-induced fibrosis.[12] They then added CCL2, TGF-ß1, and IL-13 to the fibrocyte cultures to determine their effect on collagen production. Both TGF-ß1 and CCL2 increased collagen I production from fibrocytes, but IL-13 did not. In contrast, IL-13 and TGF-ß1 both stimulated the production of collagen I by fibroblasts, but CCL2 had no effect.[12]

In another study by the same group, fibrocytes were recruited to FITC-injured lung in CCL2 knockout mice, unlike CCR2 knockout mice, implying that CCL2 was not essential for fibrocyte trafficking to the FITC injured lung. These findings were further demonstrated by blocking CCL2 activity, which did not significantly prevent fibrosis.[29]

However, neutralization of the CCR2 ligand, CCL12, significantly reduced fibrosis in the FITC mouse model of lung injury.[29] The authors suggested that CCL12 is the CCR2 ligand that promotes fibrosis in the mouse lung. However, CCL12 may only be relevant to mouse biology, as no CCL12 human homolog has been found.

Nevertheless, the CXCR4/CXCL12 biological axis appears to be the most significant chemokine target to prevent human fibrocyte recruitment in idiopathic pulmonary fibrosis, as blocking CXCL12 did not influence the migration of other leukocyte populations to the lungs of animals exposed to bleomycin, but decreased fibrosis in the lungs of animals exposed to bleomycin by attenuating the recruitment of fibrocytes.[3]

Fibrocytes in Asthma

Repair and Remodeling of the Airway in Asthma

Inflammation and airway remodeling are the hallmarks of asthma, and both contribute to disease persistence and progression.[30–33] The remodeling of asthmatic airways results in subepithelial fibrosis, mucus metaplasia, hyperplasia of myofibroblasts, hyperplasia and hypertrophy of myocytes and hypertrophy of airway epithelial cells.[32,33] The mechanisms of remodeling are not well understood but a number of cytokines and inflammatory mediators have been implicated.[32] Overexpression of the epidermal growth factor receptor (EGFR) in asthmatic airways correlates with neutrophil infiltration, IL-8 levels and mucus metaplasia.[32] Injury to the epithelium also results in the release of fibroproliferative growth factors, such as TGF-β, which results in proliferation of the underlying fibroblasts and their differentiation to myofibroblasts.[32] Other factors such as PDGF, IL-1, TNF, IL-5, IGF-1, endothelin-1, tryptases and leukotrienes have all been shown to have an effect on fibroblast proliferation and matrix production.[32] Metalloproteinases and their inhibitors, in particular MMP-9 and TIMP-1, are also involved in matrix remodeling.[32] The presumed communication between the epithelium and mesenchymal cells recapitulates that seen in embryonic lung development and argues that the remodeling in the asthmatic airway occurs from an epithelial-mesenchymal trophic unit. Identifying and characterizing the defects in proliferation and differentiation of the progenitor epithelial cells involved in the regeneration of the pulmonary epithelium in asthma has important therapeutic implications, as modifying the abnormal regeneration of these cells could provide novel treatments for asthma. In this regard, furthering our understanding of the potential contribution of mesenchymal progenitor cells to these processes may aid in the development of specific therapeutic interventions.

Fibrocytes in Airway Remodeling in Asthma

Schmidt and colleagues have shown that fibrocyte-like cells exist in the airway of asthmatic patients, and increase in number after antigen challenge and appear to differentiate into collagen producing myofibroblasts.[6] They showed that there are significant numbers of $CD34^+$, Col I^+ cells as well as a few $CD34^+$, α-SMA^+ cells below the basement membrane in the bronchial mucosa of asthmatic patients and that these cells increase dramatically at 24 hours after allergen exposure.[6] They subsequently used systematic sensitization of BALB/c mice with OVA as a mouse model of asthma, which has been shown to result in airway remodeling characterized by thickening of the lamina propria by subepithelial zone deposition of fibronectin and collagen.[6] In this mouse model, $CD34^+$, Col I^+ cells were also seen as were $CD34^+$ and α-SMA^+ cells. They then isolated $CD34^+$, Col I^+ cells from mouse peripheral blood and stained the cells with intravital PKH-26 fluorescent dye. These labeled cells were injected into the tail vein of chronically OVA exposed BALB/c mice and flow cytometric analysis of single cell suspensions from the airway tissue of these mice showed a significant increase in the number of fibrocytes in the airway submucosa compared to PBS injected controls. These PKH-26 labeled cells in the airway submucosa also showed loss of CD34 expression and an increase in α-SMA expression.[6] Finally they showed that cultures of human $CD34^+$,

Col I^+ cells in serum free medium constitutively express α-SMA; and the addition of ET-1 and TGF-β1 resulted in an increase in fibronectin and collagen III in the culture medium.[6] These findings support the notion that circulating fibrocytes may contribute to airway remodeling in asthma.

Fibrocytes in Pulmonary Vascular Remodeling

Chronic pulmonary hypertension is characterized by adventitial cell proliferation, vascular wall ECM deposition and expansion of myofibroblasts in the large and small pulmonary arteries. This process has now been shown to occur not only from resident fibroblasts, but also from circulating mesenchymal progenitor cells.[34] However, there is some controversy as to whether the circulating cells described by Frid and colleagues are true fibrocytes,[34] as they describe cells that are monocyte/macrophage-like with $CD11b^+$, $CD13^+$ and $CD14^+$ surface markers, whereas classical fibrocytes are $CD11b^+$, $CD13^+$ and CD14 negative.[1,3,34] In chronically hypoxic neonatal rat and calf models, they found that cells in the adventitia co-expressed CD45, Col I, CD14, CD11b, and collagen-prolyl-hydroxylase-α.[34] Cells in the adventitia also co-expressed CD45 and α-SMA, as well as CD68 and α-SMA.[34] To label the circulating monocytes, DiI liposomes were injected into the circulation of rats. DiI liposomes were taken up by phagocytosis and then incorporated into the cell membrane.[34] DiI labeled cells were found in the pulmonary artery adventitia of chronically hypoxic rats and not in control normoxic rats.[34] The majority of these labeled cells also expressed CD11b, ED1 and ED2, and many co-expressed collagen I. These studies highlight the potential major importance of circulating fibrocytes or fibrocyte-like cells in vascular adventitial thickening in pulmonary hypertension, as resident adventitial fibroblasts played a much smaller role in the changes seen in the vascular adventitia in pulmonary hypertension. These findings also support that targeting these fibrocyte-like cells may have therapeutic implications in the treatment of pulmonary hypertension.

Conclusions

In the last 10 years, our understanding of fibrocytes and their role in promoting fibrosis has greatly increased, but there still remains much work to be done. For example, what are all the signals involved in the recruitment of fibrocytes, are these signals different in humans and mice; what are the factors involved in their plasticity/differentiation; and what is the role of the microniche in promoting fibrocyte progression to a differentiated myofibroblast? All of these questions are critical to our understanding of fibrosis, and need to be addressed in order to design therapeutic strategies to attenuate fibrocyte biology and prevent their contribution to fibrotic disorders in organs such as the lungs.

References

1. Bucala R, Spiegel LA, Chesney J, Hogan M, Cerami A. (1994) Circulating fibrocytes define a new leukocyte subpopulation that mediates tissue repair. *Mol Med* **1**(1): 71–81.
2. Abe R, Donnelly SC, Peng T, *et al.* (2001) Peripheral blood fibrocytes: differentiation pathway and migration to wound sites. *J Immunol* **166**(12): 7556–7562.
3. Phillips RJ, Burdick MD, Hong K, *et al.* (2004) Circulating fibrocytes traffic to the lungs in response to CXCL12 and mediate fibrosis. *J Clin Invest* **114**(3): 438–446.
4. Metz CN. (2003) Fibrocytes: A unique cell population implicated in wound healing. *Cell Mol Life Sci* **60**(7): 1342–1350.
5. Quan TE, Cowper S, Wu SP, *et al.* (2004) Circulating fibrocytes: collagen-secreting cells of the peripheral blood. *Int J Biochem Cell Biol* **36**(4): 598–606.
6. Schmidt M, Sun G, Stacey MA, *et al.* (2003) Identification of circulating fibrocytes as precursors of bronchial myofibroblasts in asthma. *J Immunol* **171**(1): 380–389.
7. Chauhan H, Abraham A, Phillips JR, *et al.* (2003) There is more than one kind of myofibroblast: analysis of CD34 expression in benign, *in situ*, and invasive breast lesions. *J Clin Pathol* **56**(4): 271–276.
8. Chesney J, Bacher M, Bender A, Bucala R. (1997) The peripheral blood fibrocyte is a potent antigen-presenting cell capable of priming naive T cells *in situ*. *Proc Natl Acad Sci USA* **94**(12): 6307–6312.

9. Hartlapp I, Abe R, Saeed RW, *et al.* (2001) Fibrocytes induce an angiogenic phenotype in cultured endothelial cells and promote angiogenesis *in vivo*. *Faseb J* **15**(12): 2215–2224.
10. Chesney J, Metz C, Stavitsky AB, *et al.* (1998) Regulated production of type I collagen and inflammatory cytokines by peripheral blood fibrocytes. *J Immunol* **160**(1): 419–425.
11. Luster AD, Alon R, von Andrian UH. (2005) Immune cell migration in inflammation: present and future therapeutic targets. *Nat Immunol* **6**(12): 1182–1190.
12. Moore BB, Kolodsick JE, Thannickal VJ, *et al.* (2005) CCR2-mediated recruitment of fibrocytes to the alveolar space after fibrotic injury. *Am J Pathol* **166**(3): 675–684.
13. Murdoch C. (2000) CXCR4: Chemokine receptor extraordinaire. *Immunol Rev* **177**: 175–184.
14. Hong KM, Burdick MD, Phillips RJ, *et al.* (2005) Characterization of human fibrocytes as circulating adipocyte progenitors and the formation of human adipose tissue in SCID mice. *Faseb J* **19**(14): 2029–2031.
15. American Thoracic Society. (2000) Idiopathic pulmonary fibrosis: Diagnosis and treatment. International consensus statement. American Thoracic Society (ATS), and the European Respiratory Society (ERS). *Am J Respir Crit Care Med* **161**(2 Pt 1): 646–664.
16. American Thoracic Society/European Respiratory Society. (2002) International Multidisciplinary Consensus Classification of the Idiopathic Interstitial Pneumonias. This joint statement of the American Thoracic Society (ATS), and the European Respiratory Society (ERS) was adopted by the ATS board of directors, June 2001 and by the ERS Executive Committee, June 2001. *Am J Respir Crit Care Med* **165**(2): 277–304.
17. Strieter RM. (2005) Pathogenesis and natural history of usual interstitial pneumonia: the whole story or the last chapter of a long novel. *Chest* **128**(5 Suppl 1): 526S–532S.
18. Flaherty KR, Travis WD, Colby TV, *et al.* (2001) Histopathologic variability in usual and nonspecific interstitial pneumonias. *Am J Respir Crit Care Med* **164**(9): 1722–1727.
19. Katzenstein AL, Zisman DA, Litzky LA, *et al.* (2002) Usual interstitial pneumonia: histologic study of biopsy and explant specimens. *Am J Surg Pathol* **26**(12): 1567–1577.
20. Monaghan H, Wells AU, Colby TV, *et al.* (2004) Prognostic implications of histologic patterns in multiple surgical lung biopsies from patients with idiopathic interstitial pneumonias. *Chest* **125**(2): 522–526.

21. Iwano M, Plieth D, Danoff TM, *et al.* (2002) Evidence that fibroblasts derive from epithelium during tissue fibrosis. *J Clin Invest* **110**(3): 341–350.
22. Fukuda Y, Ishizaki M, Masuda Y, *et al.* (1987) The role of intraalveolar fibrosis in the process of pulmonary structural remodeling in patients with diffuse alveolar damage. *Am J Pathol* **126**(1): 171–182.
23. Marshall R, Bellingan G, Laurent G. (1998) The acute respiratory distress syndrome: Fibrosis in the fast lane. *Thorax* **53**(10): 815–817.
24. Kalluri R, Neilson EG. (2003) Epithelial-mesenchymal transition and its implications for fibrosis. *J Clin Invest* **112**(12): 1776–1784.
25. Hashimoto N, Jin H, Liu T, *et al.* (2004) Bone marrow-derived progenitor cells in pulmonary fibrosis. *J Clin Invest* **113**(2): 243–252.
26. Mori L, Bellini A, Stacey MA, *et al.* (2005) Fibrocytes contribute to the myofibroblast population in wounded skin and originate from the bone marrow. *Exp Cell Res* **304**(1): 81–90.
27. Epperly MW, Guo H, Gretton JE, Greenberger JS. (2003) Bone marrow origin of myofibroblasts in irradiation pulmonary fibrosis. *Am J Respir Cell Mol Biol* **29**(2): 213–224.
28. Okuma T, Terasaki Y, Kaikita K, *et al.* (2004) C-C chemokine receptor 2 (CCR2) deficiency improves bleomycin-induced pulmonary fibrosis by attenuation of both macrophage infiltration and production of macrophage-derived matrix metalloproteinases. *J Pathol* **204**(5): 594–604.
29. Moore BB, Murray L, Das A, *et al.* (2006) The role of CCL12 in the recruitment of fibrocytes and lung fibrosis. *Am J Respir Cell Mol Biol* (March 16).
30. Bousquet J, Chanez P, Lacoste JY, *et al.* (1992) Asthma: a disease remodeling the airways. *Allergy* **47**(1): 3–11.
31. Bousquet J, Jeffery PK, Busse WW, *et al.* (2000) Asthma. From bronchoconstriction to airways inflammation and remodeling. *Am J Respir Crit Care Med* **161**(5): 1720–1745.
32. Elias JA, Zhu Z, Chupp G, Homer RJ. (1999) Airway remodeling in asthma. *J Clin Invest* **104**(8): 1001–1006.
33. Fish JE, Peters SP. (1999) Airway remodeling and persistent airway obstruction in asthma. *J Allergy Clin Immunol* **104**(3 Pt 1): 509–516.
34. Frid MG, Brunetti JA, Burke DL, *et al.* (2006) Hypoxia-induced pulmonary vascular remodeling requires recruitment of circulating mesenchymal precursors of a monocyte/macrophage lineage. *Am J Pathol* **168**(2): 659–669.

Chapter 9

Role of Fibrocytes in Renal Fibrosis

Norihiko Sakai[†], Takashi Wada[*,†], Kouji Matsushima[‡] and Shuichi Kaneko[†]

Fibrosis is a characteristic pathological feature of progressive organ diseases, resulting in organ failure. Renal fibrosis is a progressive and potentially lethal disease caused by diverse clinical entities. The degree of renal fibrosis well correlates with the prognosis of renal diseases independent of their etiologies. Fibrocytes are peculiar circulating cells that share markers of leukocytes as well as mesenchymal cells. A considerable number of fibrocytes dual positive for CD45 and type I collagen or CD34 and type I collagen infiltrated the interstitium along with progression of fibrosis in an experimental murine renal fibrosis model. Most fibrocytes in the kidneys were positive for CCR7. In addition, a ligand for CCR7, secondary

*Correspondence: Takashi Wada, M.D., PhD. Disease Control and Homeostasis, Kanazawa University Graduate School of Medical Science, 13-1 Takara-machi, Kanazawa 920-8641, Japan, Tel.: +81-76-265-2000 (ext 7270); Fax: +81-76-234-4250. E-mail: twada@m-kanazawa.jp
†Disease Control and Homeostasis, Kanazawa University Graduate School of Medical Science, Ishikawa
‡Department of Molecular Preventive Medicine, Graduate School of Medicine, The University of Tokyo, Tokyo, Japan

lymphoid tissue chemokine (SLC/CCL21) co-localized with high endothelial venule-like vessels in fibrotic kidneys. CCL21/CCR7 blockade reduced the number of infiltrating fibrocytes as well as the extent of renal fibrosis, which was confirmed by a decrease in renal transcripts of pro $\alpha 1$ chain of type I collagen and transforming growth factor-β_1. These findings suggest that fibrocytes contribute to the progressive renal fibrosis dependent on CCL21/CCR7 signaling.

Introduction

Fibrosis is a characteristic pathological finding of progressive organ diseases, resulting in organ failure. Renal fibrosis is a progressive and potentially lethal disease caused by diverse clinical entities.[1,2] In addition, the degree of renal fibrosis correlates well with the prognosis of renal diseases independent of their etiologies.[3,4] The histological picture of renal fibrosis is characterized by tubular atrophy and dilation, interstitial leukocyte infiltration, accumulation of fibroblasts, and increased interstitial matrix deposition.[5] Currently, resident fibroblasts, epithelial-mesenchymal transition (EMT)-derived fibroblasts/myofibroblasts, and monocytes/macrophages are thought to be major participants in the pathogenesis of renal fibrosis.[6–9]

In addition to these cells involved, a circulating bone marrow-derived population of fibroblast-like cells (termed fibrocytes), first identified a decade ago, is thought to be another candidate that participates in organ fibrosis.[10] Fibrocytes comprise a minor fraction of the circulating pool of leukocytes (less than 1%) and share the markers of leukocytes (e.g. CD45, CD34) as well as mesenchymal cells (e.g. type I collagen, fibronectin).[11,12] Fibrocytes are present in experimental fibrosis associated with conditions such as pulmonary fibrosis, bronchial asthma, and skin wounds.[13–15] Furthermore, fibrocytes are detected in human fibrosing diseases, including nephrogenic fibrosing dermopathy and burns.[16,17] Of note, fibrocytes express chemokine receptors such as CCR7, CXCR4 and CCR2.[12,13] Recent studies demonstrate that chemokine/chemokine receptor systems on fibrocytes are involved in the recruitment of circulating fibrocytes to the sites of fibrosis.[12,13]

Here we review the pathophysiological roles of fibrocytes in renal fibrosis and discuss their trafficking into diseased kidneys from the circulation.

Fibrocytes in an Experimental Renal Fibrosis Model

1) *Presence of fibrocytes in fibrotic kidneys*

One of the unique characteristics of fibrocytes is the simultaneous expression of both leukocyte markers, such as CD45 and CD34, and type I collagen.[12] Renal fibrosis induced by unilateral ureteral obstruction (UUO) is a well-known renal fibrosis model in mice. Recently, we have revealed that fibrocytes dual positive for CD45 and type I collagen or CD34 and type I collagen were present in the interstitium, especially of the corticomedullary regions in wild-type mice fibrotic kidneys after UUO (Figs. 1a–c).[18] In addition, the number of infiltrating fibrocytes increased with the progression of fibrosis after a ureteral ligation (Fig. 1d). Thus far, it has been reported that expressions of certain chemokine receptors, such as CCR7, CXCR4 and CCR2, are detectable on fibrocytes isolated from humans and mice.[12,13] These findings prompted us to perform flow cytometry analyzes to characterize the infiltrating fibrocytes based on expressions of chemokine receptors. In wild-type mice, 37.8% of the infiltrating fibrocytes expressed CCR7 following ureteral ligation.[18] Among these CCR7-expressing fibrocytes, 66.5% of cells were positive for both CXCR4 and CCR2, and 21.1% of cells were positive for either CXCR4 or CCR2.[18]

2) *CCL21/CCR7 signaling regulates fibrocyte infiltration and renal fibrosis*

Secondary lymphoid tissue chemokine (SLC/CCL21), a ligand for CCR7, is a member of the CC chemokine family. Its first two cysteine residues are adjacent to each other. CCL21 contains six cysteines and is a potent chemoattractant for T cells, B cells and

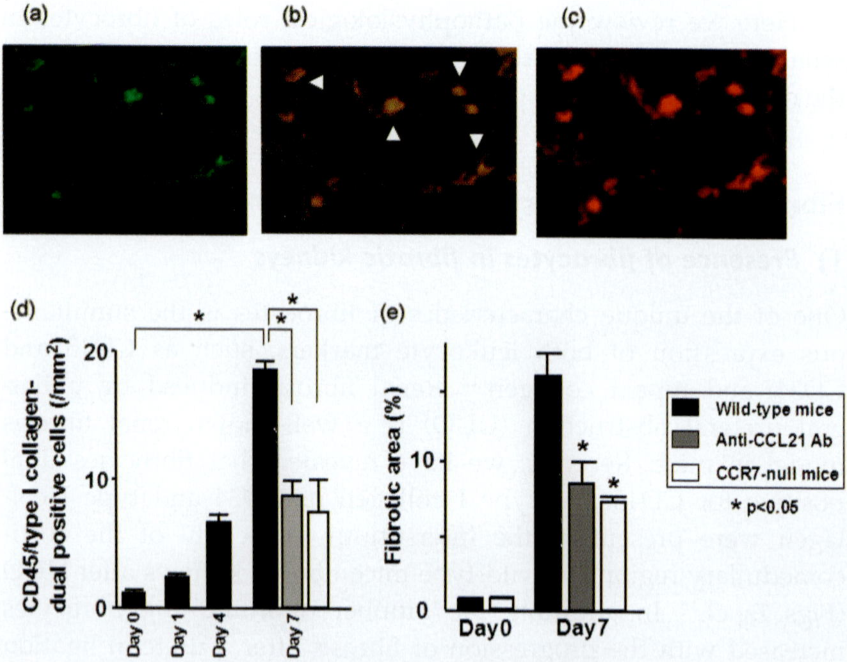

Fig. 1. Fibrocytes infiltrated the kidney after ureteral ligation an contributed to renal fibrosis. In wild-type mice, CD45- and type I collagen-dual positive fibrocytes infiltrated the interstitium, especially the corticomedullary regions after ureteral ligation (**a**: CD45; **b**: merge; **c**: type I collagen, arrowheads; CD45- and type I collagen-dual positive fibrocytes; $CD45^+/ColI^+$). The number of infiltrating fibrocytes dual positive for CD45 and type I collagen was reduced in mice treated with anti-CCL21 antibodies and in CCR7-null mice compared with that in wild-type mice 7 days after ureteral ligation (**d**). In addition, CCL21/CCR7 signaling reduced the mean interstitial fibrosis 7 days after ureteral ligation (**e**). Values are the mean ±SEM.

dendritic cells.[19–21] In addition, CCL21 also acts as a chemotactic stimulus for fibrocytes.[15] Our study demonstrated that treatment with anti-CCL21 antibodies or CCR7 deficiency in gene-targeted mice resulted in over 50% reduction in the number of CD45- and type I collagen-dual positive fibrocytes (Fig. 1d).[18] It was also noted that the number of CCR7-expressing fibrocytes was decreased in mice treated with anti-CCL21 antibodies compared with that in wild-type mice, 7 days after UUO. Based on these findings, CCL21/CCR7 signaling is thought to be the major pathway attracting fibrocytes into the kidney in this particular model. Furthermore, the extent of renal fibrosis estimated by computer-assisted measurement as well as the amount of hydroxyproline was reduced by 50% in mice treated with anti-CCL21 antibodies and in CCR7-null mice compared with those in wild-type mice 7 days after UUO (Fig. 1e).[18] Ureteral ligation enhanced the expression of pro-$\alpha 1$ chain of type I collagen (COL1A1) mRNA as well as the transforming growth factor (TGF)-β_1 mRNA in wild-type mice, which were significantly reduced by blockade of CCL21/CCR7 signaling.[18] These findings suggest that fibrocytes contribute to renal fibrosis by the production of type I collagen and that this process requires CCL21/CCR7 signaling. In contrast, the infiltration of CXCR4-positive fibrocytes was not reduced by the blockade of CCL21/CCR7.[18] In this aspect, CXCR4-positive fibrocytes are reported to migrate in response to CXCL12, a ligand for CXCR4, and traffick to the lungs in a murine model of bleomycin-induced pulmonary fibrosis.[13] Further, treatment of bleomycin-exposed animals with specific neutralizing anti-CXCL12 antibodies inhibited infiltration of CXCR4-positive fibrocytes and attenuated lung fibrosis.[13] Therefore, these findings suggest that other chemokine/chemokine receptor pathways may also be involved in the recruitment and activation of fibrocytes, resulting in progressive fibrosis. Further studies will be required to elucidate the precise mechanisms of fibrocyte trafficking into fibrotic kidneys.

3) *Infiltration routes of fibrocytes to fibrotic kidneys*

High endothelial venules (HEVs) are specialized venules that allow rapid and selective lymphocyte trafficking from the blood into the lymph nodes and Peyer's patches under physiological conditions.[22] HEVs express certain chemokines, such as CCL21[20] and EBI1-ligand chemokine/CCL19,[23] that can activate CCR7-expressing cells. In contrast, HEV-like vessels, which are observed in chronically inflamed nonlymphoid tissues, are thought to play an important role in the pathogenesis of various inflammatory diseases, such as rheumatoid arthritis and Graves' disease.[24,25]

In addition, CCL21-positive HEV-like vessels were found in synovial tissues from patients with rheumatoid arthritis.[26] With regard to human kidney diseases, HEV-like vessels are found at the corticomedullary junction and associated with interstitial leukocyte infiltration in human glomerulonephritis, whereas HEV-like vessels are not detected in normal kidneys.[27] We observed that the expression of CCL21 mRNA in diseased kidneys was upregulated with the progression of fibrosis in wild-type mice after ureteral ligation.[18] Furthermore, CCL21 protein co-localized with HEV-like vessels in the corticomedullary regions in immunohistochemical studies. The increase in the number of CCL21-positive HEV-like vessels correlated with the progression of fibrosis after ureteral ligation. It was also noted that the number of infiltrating CCR7-positive fibrocytes was markedly reduced by the blockade of CCL21/CCR7 signaling. Taken together, these findings suggest that CCR7-expressing circulating fibrocytes infiltrate the kidney via CCL21-positive HEV-like vessels as illustrated in Fig. 2, resulting in renal fibrosis.

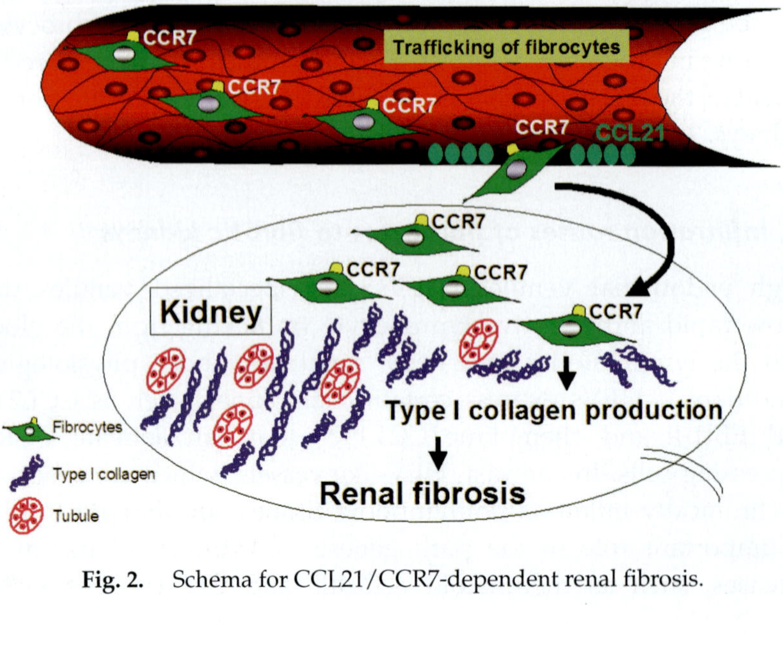

Fig. 2. Schema for CCL21/CCR7-dependent renal fibrosis.

4) *Effect of blockade of CCL21/CCR7 signaling on expression of renal monocyte chemoattractant protein-1 (MCP-1/CCL2) and infiltration of F4/80-positive macrophages*

Progressive organ fibrosis is pathologically characterized by the presence of infiltrating macrophages and accumulation of extracellular matrix (ECM), including type I collagen.[1] Currently, macrophages are thought to be involved in the development of fibrosis by secreting various cytokines and growth factors, including TGF-β_1.[28] Furthermore, recent studies reported that the CCL2/CCR2 signaling pathway is involved in the progression of fibrosis through the recruitment and activation of macrophages in various fibrotic diseases.[9,29–34] CCL2 is reported to be produced by tubular epithelial cells and infiltrating cells in fibrotic kidneys.[31] Recently, the expression of CCL2 mRNA was shown to be enhanced in fibrocytes under fibrotic circumstances.[11] In addition, we observed that renal expression of CCL2 mRNA and the infiltration of F4/80-positive macrophages as well as CCR7-expressing fibrocytes were significantly reduced in mice treated with anti-CCL21 antibodies and in CCR7-null mice after ureteral ligation compared with those in UUO-treated wild-type mice.[18] Our previous reports demonstrated that monocytes/macrophages also contribute to renal fibrosis since the blockade of CCL2/CCR2 signaling resulted in a 30% reduction of renal fibrosis after ureteral ligation.[9,33] In contrast, fibrosis and infiltration of fibrocytes in the kidneys was reduced up to 50% by the inhibition of CCL21/CCR7 signaling.[18] Taken together, these findings suggest that CCR7-expressing fibrocytes are involved in the pathogenesis of fibrosis not only by secreting collagen but also by regulating the infiltration and activation of macrophages through CCL2 production.

Fibrocytes in Human Renal Diseases

Thus far, the precise role of fibrocytes in the pathogenesis of human renal diseases is still unclear. CD34-positive spindle cells are reported to be present in the interstitium in patients with glomerulonephritis.[35] In addition, the density of CD34-positive spindle cells showed a positive correlation with the interstitial volume, whereas that was not related to the kidney function parameters, such as serum creatinine and urea.[35] Circulating fibrocytes express CD34, whereas expression of CD34 by fibrocytes decreases over time under certain conditions.[14,15] TGF-β has been reported to induce a decrease in cell surface CD34 and an increase in α-smooth muscle actin, which is a characteristic marker of contractile myofibroblasts.[14,15] In contrast, additional cell surface markers, such as CD45, have been reported to be stably expressed on fibrocytes.[18] Thus far, we observed that fibrocytes dual positive for CD45 and type I collagen were present in the interstitium in patients with various renal diseases and that the number of infiltrating fibrocytes correlated well with the degree of renal fibrosis and renal function (unpublished data). Therefore, it is suggested that fibrocytes may be involved in the progression of human renal diseases, especially fibrotic lesions. Further studies will be needed to elucidate the precise roles of fibrocytes in human renal diseases.

Concluding Remarks

In summary, fibrocytes are novel collagen-producing cells and contribute to the progressive renal fibrosis, dependent on CCL21/CCR7 signaling. Regulating the recruitment and activation of fibrocytes may be a novel therapeutic strategy for renal fibrosis.

References

1. Wada T, Razzaque MS, Matsushima K, *et al.* (2004) Pathological significance of renal expression of proinflammatory molecules. In: Razzaque MS (ed.), *Fibrogenesis: Cellular and Molecular Basis*. Landes Bioscience Eurekah, Georgetown, Texas, pp. 9–26.
2. Bohle A, Muller G.A, Wehrmann M, *et al.* (1996) Pathogenesis of chronic renal failure in the primary glomerulopathies, renal vasculopathies, and chronic interstitial nephritides. *Kidney Int* **Suppl 49**: S2–S9.

3. Risdon R-A, Sloper J-C, de Wardener H-E. (1968) Relationship between renal function an histologic changes found in renal-biopsy specimens from patients with persistent glomerulonephritis. *Lancet* **2**: 363–366.
4. Nath K-A. (1998) The tubulointerstitium in progressive renal disease. *Kidney Int* **54**: 992–994.
5. Vielhauer V, Anders H-J, Mack M, *et al.* (2001) Obstructive nephropathy in the mouse: Progressive fibrosis correlates with tubulointerstitial chemokine expression and accumulation of CC chemokine receptor 2- and 5-positive leukocytes. *J Am Soc Nephrol* **12**: 1173–1187.
6. Strutz F, Zeisberg M, Ziyadeh F-N, *et al.* (2002) Role of basic fibroblast growth factor-2 in epithelial-mesenchymal transformation. *Kidney Int* **61**: 1714–1728.
7. Zeisberg M, Hanai J, Sugimoto H, *et al.* (2003) BMP-7 counteracts TGF-β1-induced epithelial-to-mesenchymal transition and reverse chronic renal injury. *Nat Med* **9**: 964–968.
8. Iwano M, Plieth D, Danoff TM, *et al.* (2002) Evidence that fibroblast derive from epithelium during tissue fibrosis. *J Clin Invest* **110**: 341–350.
9. Kitagawa K, Wada T, Furuichi K, *et al.* (2004) Blockade of CCR2 ameliorates progressive fibrosis in kidney. *Am J Pathol* **165**: 237–246.
10. Bucala R, Spiegel L-A, Chesney J, *et al.* (1994) Circulating fibrocytes define a new leukocyte subpopulation that mediates tissue repair. *Mol Med* **1**: 71–81.
11. Chesney J, Metz C, Stavitsky A-B, *et al.* (1998) Regulated production of type I collagen and inflammatory cytokines by peripheral blood fibrocytes. *J Immunol* **160**: 419–425.
12. Moore B-B, Kolodsick J-E, Thannickal V-J, *et al.* (2005) CCR2-mediated recruitment of fibrocytes to the alveolar space after fibrotic injury. *Am J Pathol* **166**: 675–684.
13. Phillips R-J, Burdick M-D, Hong K, *et al.* (2004) Circulating fibrocytes traffic to the lungs in response to CXCL12 and mediate fibrosis. *J Clin Invest* **114**: 438–446.
14. Schmidt M, Sun G, Stacey M-A, *et al.* (2003) Identification of circulating fibrocytes as precursors of bronchial myofibroblasts in asthma. *J Immunol* **171**: 380–389.
15. Abe R, Donnelly S-C, Peng T, *et al.* (2001) Peripheral blood fibrocytes: Differentiation pathway and migration to wound sites. *J Immunol* **166**: 7556–7562.

16. Hauser C, Kaya G, Chizzolini C. (2004) Nephrogenic fibrosing dermopathy in a renal transplant recipient with tubulointerstitial nephritis and uveitis. *Dermatology* **209**: 50–52.
17. Yang L, Scott P-G, Giuffre J, *et al.* (2002) Peripheral blood fibrocytes from burn patients: Identification and quantification of fibrocytes in adherent cells cultured from peripheral blood mononuclear cells. *Lab Invest* **82**: 1183–1192.
18. Sakai N, Wada T, Yokoyama H, *et al.* (2006) Secondary lymphoid tissue chemokine (SLC/CCL21)/CCR7 signaling regulates fibrocytes in renal fibrosis. *Proc Natl Acad Sci USA*, **103**: 14098–14103.
19. Campbell J-J, Bowman E-P, Murphy K, *et al.* (1998) 6-C-kine (SLC), a lymphocyte adhesion-triggering chemokine expressed by high endothelium, is an agonist for the MIP-3β receptor CCR7. *J Cell Biol* **141**: 1053–1059.
20. Gunn M-D, Tangemann D-K, Tam C, *et al.* (1998) A chemokine expressed in lymphoid high endothelial venules promotes the adhesion and chemotaxis of naïve T lymphocytes. *Proc Natl Acad Sci USA* **95**: 258–263.
21. Ogata M, Zang Y, Wang Y, *et al.* (1999) Chemotactic response toward chemokines and its regulation by transforming growth factor-β1 of murine bone marrow hematopoietic progenitor cell-derived different subset of dendritic cells. *Blood* **93**: 3225–3232.
22. Kraal G, Mebius R-E. (1997) High endothelial venules: Lymphatic traffic control and controlled traffic. *Adv Immunol* **65**: 347–395.
23. Baekkevold E-S, Yamanaka T, Palframan R-T, *et al.* (2001) The CCR7 ligand ELC (CCL19) is translocated in high endothelial venules and mediates T cell recruitment. *J Exp Med* **193**: 1105–1112.
24. Dinther-Janssen A-C-H-M, Pals S-T, Scheper R, *et al.* (1990) Dendritic cells and high endothelial venules in the rheumatoid synovial membrane. *J Rheumatol* **17**: 11–17.
25. Kabel P-J, Voorbij H-A-M, Haan-Meulman M, *et al.* (1989) High endothelial venules present in lymphoid cell accumulations in thyroids affected by autoimmune disease: A study in men and BB rats of functional activity and development. *J Clin Endocrinol Metab* **68**: 744–751.
26. Weninger W, Carlsen H-S, Goodarzi M, *et al.* (2003) Naïve T cell recruitment to nonlymphoid tissues: A role for endothelium-expressed CC

chemokine ligand 21 in autoimmune disease and lymphoid neogenesis. *J Immunol* **170**: 4638–4648.

27. Segawa C, Wada T, Takaeda M, *et al.* (1997) *In situ* expression and soluble form of P-selectin in human glomerulonephritis. *Kidney Int* **52**: 1054–1063.
28. Border W-A, Noble N-A. (1994) Transforming growth factor beta in tissue fibrosis. *N Engl J Med* **331**: 1286–1292.
29. Suga M, Iyonaga K, Ichiyasu H, *et al.* (1999) Clinical significance of MCP-1 levels in BALF and serum in patients with interstitial lung diseases. *Eur Respir J* **14**: 376–382.
30. Lehmann M-H, Kuhnert H, Muller S, Sigusch H-H. (1998) Monocyte chemoattractant protein 1 (MCP-1) gene expression in dilated cardiomyopathy. *Cytokine* **10**: 739–746.
31. Wada T, Furuichi K, Segawa C, *et al.* (1999) MIP-1α and MCP-1 contribute to crescents and interstitial lesions in human crescentic glomerulonephritis. *Kidney Int* **56**: 995–1003.
32. Moore B-B, Paine R-III, Christensen P-J, *et al.* (2001) Protection from pulmonary fibrosis in the absence of CCR2 signaling. *J Immunol* **167**: 4368–4377.
33. Wada T, Furuichi K, Sakai N, *et al.* (2004) Gene therapy via blockade of monocyte chemoattractant protein-1 for renal fibrosis. *J Am Soc Nephrol* **15**: 940–948.
34. Wada T, Yokoyama H, Furuichi K, *et al.* (1996) Intervention of crescentic glomerulonephritis by antibodies to monocyte chemotactic and activating factor (MCAF/MCP-1). *FASEB J* **12**: 1418–1425.
35. Okon K, Szumera A, Kuzniewski M. (2003) Are CD34+ cells found in renal interstitial fibrosis? *Am J Nephrol* **23**: 409–414.

Chapter 10

Role of Fibrocytes in Atherogenesis

Heather Medbury

Introduction

Atherosclerosis is an inflammatory disease characterized by the formation of lesions in large and medium sized arteries supplying the heart, brain or extremities.[1] Multiple lesions may be present and they are primarily located in regions exposed to mechanical forces, in particular low shear stress.[2] The disease progresses slowly, starting in early childhood with the formation of a fatty streak that develops into a complex plaque generally several decades later. By early adulthood most individuals in developed countries will have some advanced plaques.[3] Disease progression can lead to various clinical outcomes, most significantly infarction causing heart attack or stroke.[1] As such atherosclerosis is a major cause of morbidity and mortality.

Atherosclerosis begins with endothelial dysfunction and accumulation of lipoproteins in the subendothelial space which promote leukocyte transmigration through the endothelial cell

Vascular Biology Research Centre, Dept Surgery, University of Sydney, Westmead Hospital, Westmead, NSW, Australia.

layer.[1,4] Monocyte-derived macrophages engulf modified lipids forming foam cells, and these along with T cells and smooth muscle cells form a fatty streak.[4] Most of the lipid at this stage is within cells, predominantly in macrophages, but some also in smooth muscle cells (SMC).[5] Progression to an intermediate lesion is associated with extracellular pools of lipid below the macrophage and foam cells layers.[5] The further accumulation of extracellular lipid leads to formation of a fatty core[5] and at this stage the lesion is defined as an *atheroma*, the first stage of the advanced lesion.[6] Overlying the core, the intima contains predominantly macrophages/macrophage foam cells with isolated smooth muscle cells and minimal collagen.[6] With time a fibrous cap develops, fencing the core off from the vessel lumen.[7] At this stage the lesion is classed as a *fibroatheroma* with several subcategories depending upon the overall composition.[6] The lesion may then progress to a complicated lesion whose main histology includes surface defects such as fissure and ulcerations, presence of a hematoma/hemorrhage and evidence of thombosis.[6]

It was proposed by Ross in 1973[8] that atherosclerosis develops as a response to injury and as such, with the development of atherosclerosis occurring over decades of life, it can be viewed as a perpetual wound. Classically, wound healing involves three phases: inflammation, tissue formation and tissue remodeling, with these phases overlapping in time.[9] Acute wounds heal in this relatively ordered fashion, whereas chronic wounds are characterized by non-resolving inflammation[10]: as is the case in atherosclerosis where the lesion remains in a largely inflammatory state (focused largely in the fatty core) with the degree of tissue formation/remodeling (fibrous cap development) varying widely between patients. Tissue remodeling may be expansive remodeling where the plaque pushes the arterial wall outwards, or constrictive remodeling that accelerates luminal narrowing.[11] Though the inflammatory response in atherosclerosis may have begun as a protective mechanisms, it fails to resolve due to continual injury converting a usual tissue injury/repair process into atherosclerosis, i.e. the persistence of both an inflammatory state and tissue hyperplasia.[8]

With the finding by Bucala[12] that fibrocytes contribute to wound healing, it can be speculated that fibrocytes may play a role in atherogenesis — in particular in formation of the fibrous cap, be it either in its initial formation or in the healing response generated upon plaque rupture. Understanding the role of fibrocytes in atherosclerosis will aid in developing approaches to promote resolution of vascular disease. This may potentially lead to a decrease in cardiovascular events such as myocardial infarction and stroke, which will not only reduce the morbidity and mortality associated with atherosclerosis but also limit the associated, health, social and economic burdens it poses.

Atherosclerosis: The Perpetual Wound

Inflammation: Fatty Core Development

The normal healthy vessel consists of three layers: the intima, media and adventitia. The intima is composed of a single layer of endothelial cells (EC) with an underlying basement membrane and elastic lamina. The media is composed of SMC and the outer adventitia is composed of fibroblasts. The endothelium plays a key role in maintaining vascular function by regulating influx of lipoproteins and leukocytes, and the release of various growth factors, such as transforming growth factor beta (TGF-β) that helps maintain vessel structure.[13] With aging some thickening of the intima is known to occur[14] and it is thought that this accumulation of SMC supports atherosclerotic development and progression.[14,15]

Endothelial injury arises from an array of factors such as hyperlipidemia (with uptake of the lipids into the subendothelial space), hypertension, diabetes, free radicals caused by cigarette smoking, and other forms of oxidative damage.[1,16] The result is increased endothelial cell permeability, development of a procoagulant state, release of vasoactive molecules, cytokines, growth factors and upregulation of adhesion molecules such as E-selectin and P-selectin promoting adhesion of platelets and leukocytes.[1] Platelets adhere to the various components of the exposed matrix or directly to intact but activated EC.[17,18] As in wound healing, platelet adhesion and activation precedes the invasion of leukocytes.[19] A range of chemokines, such as MCP-1, are released which are potent chemoattractants for monocytes.[20,21] The monocytes migrate into the vessel wall where they are transformed into macrophages and upon internalizing the modified/oxidized low density lipoprotein, become foam cells.[16,22] The inflammatory phase is mediated predominantly by macrophages and T cell subsets[1] with a continual influx of these cells. Their activation leads to release of further cytokines such as IL-1, TNF-α IFN-γ and IL-12, chemokines and growth factors, such as TGF-β and PDGF. The inflammatory cytokines induce macrophage activation and eventually foam cell death and thus the formation of an acellular lipid core.[1,16,23] If the injurious agent(s) are not removed, the initial protective response becomes an injurious response with persistent inflammation, inducing a fibrotic response which itself, when excessive, becomes part of the disease process.[1]

Tissue Formation/Remodeling: Development of the Fibrous Cap

The fibrous cap forms as a further protective response to the injury, fencing the fatty core off from the vessel lumen.[1,7] The propagation of new tissue formation in a wound, or in this case atherosclerosis, is initiated largely by activated macrophages and to some extent T cells and endothelial cells that release a range of growth factors such as TGF-β and PDGF.[24,25] Interestingly, even foam cells themselves are stimulated by oxysterols to promote TGF-β synthesis.[26] Macrophages secrete urokinase-type plasminogen activator (uPA), a serine proteinase that converts plasminogen to plasmin, which activates TGF-β[27]; TGF-β promotes cell differentiation and production of various matrix components including collagen.[28] So too PDGF promotes cell differentiation,[29] migration and proliferation. In addition, once plaques have reached an advanced stage there is usually focal deposition of platelets on the lumen which are thought to stimulate smooth muscle cell growth.[3]

Traditionally, it has been thought that the fibrous cap forms from the migration of medial SMC through the internal elastic lamina to the intima with proliferation and concomitant production of collagen.[4,8] While this proposal remains the accepted hypothesis, there are now several anomalies that cast doubt on this process. Firstly, SMC proliferation in atherosclerosis is limited[30] and even after plaque rupture, where cell proliferation is increased, these cells have been identified as macrophages (based on CD68 expression), not SMC.[31] Secondly, the plaque smooth muscle cells have several distinct characteristics compared to the medial smooth muscle.[32] Even more compelling is *in vivo* data by Sata[33] demonstrating accumulation of bone marrow-derived cells in the fibrous cap of plaques formed in the apo E−/− mouse model, with most of these cells expressing α-SMA. Similarly, it has been shown in human coronary atherosclerosis that after bone marrow transplantation, SMC of donor origin are found in the plaque.[34] The exact phenotype of the transplantation-derived SMC was not determined in these studies with various circulating progenitors suggested as possibilities[34,35] — indeed SMC progenitor cells have been found in human blood.[29] We proposed that these cells may be fibrocytes since fibrocytes acquire SMC characteristics such as a spindle-shape,[12] expression of α-SMA[36] and production of extracellular matrix proteins such as collagen I and III.[12] Consistent with Sata's findings, fibrocytes have been shown to differentiate from a circulating precursor that is bone marrow-derived.[37,38] Notably, fibrocyte differentiation is promoted by TGF-β,[36,38] a factor that is associated with plaque stability and thus increased α-SMA positive cells in the cap[39] as well as by fibronectin which is prominent in the sub-endothelium of developing fibrous plaques — creating a scaffold for the organization and construction of new tissue.[40,41] Fibrocytes also express CCR2,[42] the ligand for MCP-1 which is a potent chemokine produced in atherosclerosis.

Fibrocytes have been identified in atherosclerosis only recently by the author's group.[43,44] This is partly because, unlike other wound models, clinical specimens of atherosclerosis are at a rather advanced stage of the disease when they are obtained. At this stage, formation of the cap is generally long established and thus fibrocytes, if they did contribute to its formation, will have differentiated to a stage where the origin of the cells is difficult to demonstrate. As such the focus in atherosclerosis has been identifying the source of the α-SMA positive cells in the cap, and while fibrocytes can express α-SMA, this is only on mature fibrocytes.[36] Fibrocyte expression of α-SMA is promoted by TGF-β stimulation, which conversely down-regulates CD34 expression[38] — a marker originally used in combination with collagen and vimentin to characterize fibrocytes.[12] The transformation of fibrocytes from $CD34^+/\alpha$-SMA^- to $CD34^-/\alpha$-SMA^+ is thought to occur quickly at the site of injury,[38] making it difficult to detect double staining for both α-SMA and CD34. The likelihood of detecting fibrocytes in clinical atherosclerosis specimens by their transient expression of CD34 alone is minimal, and this is in fact the case with the detection of appreciable numbers of CD34 positive cells in the intima (other than on endothelial cells), an uncommon occurrence.[44] The detection of cells co-expressing CD34 with α-SMA is therefore even more unlikely, though the authors have detected it in a few specimens.[44]

Staining directly for collagen I expression in histological specimens is unsuitable, as it is an extracellular protein making it difficult to determine which cell actually produced it. However, collagen I is secreted into the extracellular space as a precursor form-procollagen which contains non-collagenous polypeptides at both the N- and C-terminal ends which are cleaved off after secretion of the molecule.[45] Thus procollagen I can be stained for in place of collagen I.

While fibrocytes were initially characterized by the expression of CD34, collagen and vimentin,[12] the minimal characteristics for the detection of fibrocytes has more recently been broadened to be the co-expression of collagen production and uniquely hematological markers.[37] In this respect, Yang[46] identified leukocyte specific protein 1 (LSP-1) as a leukocyte marker retained on fibrocytes after the down-regulation of CD34. Thus histologically, the combination of procollagen 1, and LSP-1 has been used to identify fibrocytes in scar tissue[46] and by the author's group in atherosclerosis[44] as outlined below.

Growth of the fibrous cap can be detected by staining for procollagen I. Not all atherosclerosis specimens stain positive for procollagen, which is indicative of not all specimens being in an active phase of growth upon specimen retrieval. Procollagen I, when present, is predominantly evident in the cap region; however, it is also evident in some samples around the core periphery, including below the fatty core. This would be indicative of expansive remodeling. Procollagen I staining is generally evident as diffuse staining, due to its secretion, though in some samples distinct cellular staining is evident. CD34, when present, is evident on spindle-shaped cells aligned approximately parallel to the luminal surface in regions staining for procollagen I. Not all procollagen I regions contain CD34 positive cells, which can be expected due to the transient expression of CD34 by fibrocytes as well as the persistent presence of procollagen in the extracellular matrix for a period before cleavage of its non-collagenous ends. As such, regions with diffuse but no distinct cellular staining for procollagen I were devoid of CD34. In those regions in which CD34 and procollagen I were co-localized, double staining of these markers was evident (Figs. 1A and B). In addition, double staining for vimentin and CD34 is also evident, and is much clearer than CD34/procollagen I staining, as both markers are cellular rather than secreted (Fig. 1C).

LSP-1 is abundant in the atherosclerotic plaques, as it is present on all infiltrating leukocytes (displaying a round morphology) and retained on foam cells (displaying an enlarged round/oblong shape) in the core periphery. It also co-localizes well with procollagen (distinct cellular as opposed to diffuse) rich regions, where it displays an elongated/spindle morphology and double staining of these markers can be seen. (However, as procollagen I is often evident around the core periphery as are foam cells (LSP-1 positive), co-localization of procollagen I and LSP-1 is not always indicative of fibrocyte formation.)

Consistent with being fibrocytes, these regions stain strongly for TGF-β and fibronectin which are two key factors identified in the differentiation of mononuclear precursors into fibrocytes.[36] Fibronectin

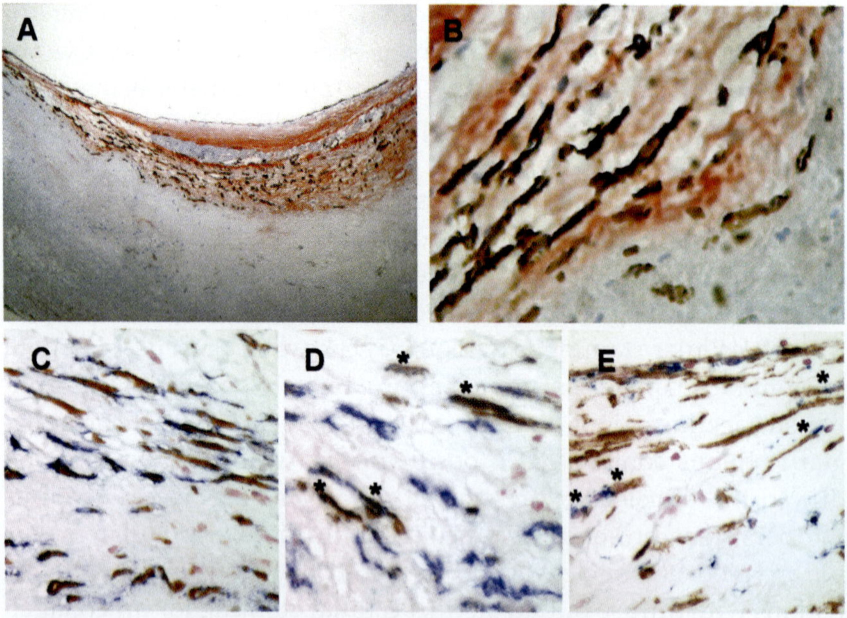

Fig. 1. A carotid endarterectomy specimen showing (**A**) CD34 (brown) and procollagen I (red). (**B**) higher magnification of (**A**) to show close up of co-expression of CD34 and pr-collagen I, (**A&B** haematoxylin counterstain). (**C**) Vimentin (brown) with CD34 (blue) — most cells are double stained. (**D**) α-SMA (brown) and CD34 (blue) with double stained cells labeled by *. (**E**) α-SMA (brown) and CD68 (blue) with double stained cells labeled by *. (**C–E** nuclear fast red counterstain).

appears in wounds prior to collagen expression and is thought to serve as a template for collagen fibril organization.[47]

The co-expression of procollagen I with CD68 (a marker which we have recently identified on fibrocytes *in vitro*)[44] was also evident in some samples reflecting their monocyte origin. The contribution of fibrocytes to the α-SMA population in the cap is indicated by α-SMA co-expression with CD34 (Fig. 1D) or CD68 (Fig. 1E). Admittedly, very few specimens contain these combinations of markers but this is to be expected as marker expression changes upon differentiation and maturation of the fibrocyte. The expression of CD68 by fibrocytes is interesting as the identification of proliferating cells in atherosclerosis pinpoints CD68 positive cells.[31] These cells have been assumed to be macrophages; however, our finding of CD68 on fibrocytes suggests that some of these cells may be fibrocytes. While this has not yet been determined in atherosclerosis, *in vivo* evidence in asthma models suggests that fibrocytes proliferate early after their appearance in the tissues.[48]

Thus fibrocytes contribute to formation of the fibrous cap — a role which was traditionally was thought to be due solely to SMC.[8] Their formation in the cap, with their expression of collagen and co-localization with the plaque-stabilizing factor TGF-β, indicates that they play a role in plaque stabilization.

Plaque Rupture

The persistent inflammatory state of the atherosclerotic plaque literally undermines the development of the fibrotic cap. The fatty core is rich in CD40L which promotes the expression of a range of proinflammatory factors, including the production of matrix metalloproteinases that degrade the various components of the extracellular matrix in the cap.[49] This thinning of the cap produces an unstable or vulnerable plaque that is prone to disruption or rupture. There are varying degrees of plaque disruption, including erosion of the endothelial layer, rupture including small breaks in the cap leading to intraplaque hemorrhage, or more significant tears/multiple tears in the cap.[3] The healing response generated results in rapid growth of the lesion[3] and is likewise thought to be due to SMC migration.[50] However, we have found that fibrocytes contribute to healing of a ruptured plaque.[44] The wound environment is conducive to their formation as clots formed upon intraplaque hemorrhage are composed largely of provisional matrix proteins such as fibrin and fibronectin. In addition, the predominant cell population in clots is platelets[3] and these release TGF-β at the site of injury[51] as well as a range of factors which are chemotactic for monocytes. We have recently shown that platelets can induce monocyte differentiation into fibrocytes *in vitro*.[44,52] Platelets have their own novel mechanisms for activating TGF-β from its latent form[53] and thus could induce fibrocyte formation in the healing plaque.

The healing of a ruptured plaque takes weeks.[50] As collagen matures, more intramolecular and intermolecular cross-links are formed which are important in giving it strength and stability over time.[51] However, scar tissue is weaker than that in the normal tissue as the collagen fibers are much smaller and have a more random appearance.[51] This situation, combined with the continued assault from matrix metalloproteinases, means that plaques may undergo cycles of rupture and repair and it is thought that in many cases, fatal ruptures represent the end stage of this process.[54] Any circumstance that prevents cap repair is potentially dangerous,[55] accordingly, interrupting the cycle of subclinical plaque rupture/erosion and repair inhibits the significant increase in plaque burden and negative remodeling that occurs.[54,55] Thus promoting plaque stabilization through fibrocyte formation would limit the likelihood of plaque rupture occurring and furthermore aid in the repair process if rupture did take place. It should be noted, however, that fibrocytes *in vitro* have been shown to express scavenger receptors (and can form fibrocyte-foam cells *in vitro* — H. Medbury unpublished observations) as well as produce matrix metalloproteinases.[37] To what degree, and under what circumstances, this behavior is exhibited *in vivo* and more importantly the effect on plaque stability is not yet known.

TGF-β: The Key Factor

TGF-β is secreted in a latent form, L-TGF-β, that cannot bind to its receptor and thus is inactive.[56] Its activation is achieved through a range of factors, including plasmin.[27] However, in atherosclerosis TGF-β levels are reduced[57] and there is diminished activity. This may occur due to elevated lipids[27] (causing sequestration of TGF-β),[58] PAI-1, or the low levels of urokinase and plasmin in atherosclerosis.[59] It is thought that the loss of TGF-β could allow the influx of leukocytes and formation of a fatty core with unstable phenotype.[13] Indeed stable plaques have more TGF-β than their unstable counterparts[60]: a fact that is clearly evident in mouse models of atherosclerosis. When TGF-β or its receptor are blocked, the lesions that form are larger and exhibit a more vulnerable phenotype with reduced collagen and conversely increased inflammation.[13,39,60] In addition, the loss of TGF-β would result in loss of fibrocyte formation (as well as a reduction in the collagen and α-SMA they produce) which would in part explain the reduced cap thickness obtained when TGF-β (or its receptor) are inhibited.

The stabilizing effect of TGF-β in atherosclerosis is affected by its multiple effect on the formation of the core and cap:

- It suppresses EC proinflammatory adhesion molecule expression,[61] thereby suppressing leukocyte recruitment.
- It is a deactivating factor for macrophages,[62] downregulating the scavenger receptors CR-A and CD36, thereby decreasing foam cell formation.[63]
- It is a deactivating factor for T cells.[64,65]
- It is associated with reduced CD40L levels.
- It drives matrix production in the fibrous cap.[39,60]
- It decreases secretion of matrix metalloproteinases while stimulating the production of tissue inhibitor of metalloprotease (TIMP).[66]
- It is involved in fibronectin deposition.[67]

Many of these actions of TGF-β are also likely to directly or indirectly influence fibrocyte formation and function in atherosclerosis. Interestingly, statins (the major pharmacological agent used in the treatment of atherosclerosis) increase TGF-β levels,[68] leading to a more stable plaque and so an additional indirect mechanism of action of statins may be promotion of fibrocyte formation.

Fibrocytes: Friend or Foe in Atherosclerosis

As presented throughout this book, fibrocytes play a detrimental role in a range of pathological conditions. However, are fibrocytes a friend or foe in atherogenesis? As outlined above, atherosclerosis is a perpetual wound with an excessive inflammatory phase accompanied by a fibrotic phase with neither of these phases adequately resolving. To complete the healing response, ideally resolution of both these phases needs to occur. So do we aim to inhibit fibrocyte formation in the treatment of atherosclerosis? Most likely not — primarily because of the importance of the fibrous cap in the clinical outcome of atherosclerosis. For though fibrocytes contribute to the growth of the lesion and thus stenosis of the vessel, the clinical significance of an atherosclerotic plaque is dependant more on its composition than its size.[69,70] In general, it is not the large plaques that cause myocardial infarcts, but smaller lesions with an unstable phenotype, as these are prone to rupture.[1] Plaques with a thick fibrous cap and a relatively small fatty core are reasonably stable,[69–71] while plaques having a relatively large fatty core (>40%) and thin fibrous cap are unstable.[70] Rupture may result in spilling out of the highly thrombogenic contents of the core, leading to thrombosis and thus blockage in the downstream vessel, inducing a heart attack or stroke. Thus, while fibrocyte formation is part of the disease process, its beneficial effects in stabilizing the plaque outweigh the increase in plaque volume that it generates.

Monocytes: A Source of Fibrocytes

Fibrocytes were initially generated *in vitro* from the culture of peripheral blood leukocytes,[12] but have more recently been shown to be derived from a $CD14^+$ enriched mononuclear population.[36,72] The transformation is induced by T cell factors[36,37] of which TGF-β is known to play a key role.[36,38] Consistent with this, the author's group has likewise shown, *in vitro*, that platelets can induce monocyte transformation into a fibrocyte, adopting the typical spindle shape and expression of the markers collagen I, CD13, CD34, CD45 and vimentin.[44,52] In addition to known fibrocyte markers, they also stain positive for CD68.[44] The platelet induced fibrocytes can also be seen to mature as indicated by up-regulation of α-SMA expression.[44] That monocytes may be transformed into a SMC like cell, is consistent with recent findings of monocytes being a source of mesenchymal progenitors[73,74] and the work by Simper[29] showing that blood mononuclear cells can be transformed into SMC. It is not entirely clear whether fibrocytes are a unique subpopulation of $CD14^+$ mononuclear cells, derived from a subpopulation of monocytes or if all monocytes can be transformed into fibrocytes. If they are a circulating subpopulation, this would have to be at a pre-collagen secretion phase as collagen activates platelets.[75] However, isolated fibrocytes do express chemokine receptors that enable them to migrate to the site of injury[36] as well as adhesion molecules.[37] Once there, the degree to which they are transformed into a SMC is not known, though they are known to be transformed into myofibroblasts in skin.[38]

If fibrocytes are derived from the same mononuclear population as monocyte-derived foam cells, then ultimately the balance between inflammation and fibrosis — which is critical in determining the stability of the plaque[39] — may hinge on the differentiation of one cell, the monocyte. Promoting monocyte transformation into fibrocytes would have a two-fold action in plaque stabilization, as it would not only result in formation of the cap, but concomitantly limit monocyte transformation into macrophages (and consequently foam cells) and thus prevent expansion of the fatty core.

Intimal Hyperplasia

When plaque growth leads to significant stenosis of the artery, surgical intervention such as endarterectomy, angioplasty (with or without stent placement) or bypass may be undertaken. This surgical wound generates a healing response that can lead to intimal hyperplasia development that is a major cause of restenosis. As with atherosclerosis, the traditional view of formation of the neointima describes the accumulation of medial SMC that have migrated, proliferated and deposited extracellular matrix.[76] Various sources of SMC have now been proposed, including their derivation from cells in the adventitia.[77] More recently, however, a number of experimental models have demonstrated the contribution of bone marrow-derived cells in the formation of the intimal hyperplasia mass.[33,35,78,79] We have recently shown that fibrocytes contribute to the α-SMA population in intimal hyperplasia, with the majority of the α-SMA positive cells in an ovine patch graft model arising from the circulating progenitor.[52]

Conclusion

Fibrocytes play a role in atherogenesis by their contribution to the formation of the fibrous cap, thereby causing plaque growth and luminal narrowing. Paradoxically, rather than being completely detrimental to the clinical outcome of atherosclerosis, their formation is important in the stabilization of the plaque — particularly through their production of collagen. Thus promoting their formation may lead to a decrease in plaque rupture and accordingly a decrease in clinical events such as myocardial infarction and stroke.

So, too, fibrocytes play a major role in the intimal hyperplasia that develops after surgical intervention of atherosclerosis. In this situation, however, their formation is detrimental as they promote restenosis of the vessel and so, in contrast to atherosclerosis, their formation needs to be inhibited. The quandary is that any attempt to inhibit their formation in intimal hyperplasia by systemic approaches may lead to destabilization of other atherosclerotic plaques within the vascular system — as patients tend to have multiple plaques. The treatment of both atherosclerosis and intimal hyperplasia, therefore, will be a complex issue and should perhaps focus primarily on the resolution of inflammation rather than fibrosis.

References

1. Ross R. (1999) Atherosclerosis — an inflammatory disease. *N Engl J Med* **340**: 115–126.
2. Traub O, Berk BC. (1998) Laminar shear stress: mechanisms by which endothelial cells transduce an atheroprotective force. *Arterioscler Thromb Vasc Biol* **18**: 677–685.
3. Davies MJ. (1996) Stability and instability: two faces of coronary atherosclerosis. The Paul Dudley White Lecture 1995. *Circulation* **94**: 2013–2020.
4. Ross R. (1995) Cell biology of atherosclerosis. *Annu Rev Physiol* **57**: 791–804.
5. Stary HC, Chandler AB, Glagov S, *et al.* (1994) A definition of initial, fatty streak, and intermediate lesions of atherosclerosis. A report from

the Committee on Vascular Lesions of the Council on Arteriosclerosis, American Heart Association. *Arterioscler Thromb* **14**: 840–856.

6. Stary HC, Chandler AB, Dinsmore RE, *et al.* (1995) A definition of advanced types of atherosclerotic lesions and a histological classification of atherosclerosis. A report from the Committee on Vascular Lesions of the Council on Arteriosclerosis, American Heart Association. *Arterioscler Thromb Vasc Biol* **15**: 1512–1531.
7. Munro JM, Cotran RS. (1988) The pathogenesis of atherosclerosis: Atherogenesis and inflammation. *Lab Invest* **58**: 249–261.
8. Ross R, Glomset JA. (1973) Atherosclerosis and the arterial smooth muscle cell: proliferation of smooth muscle is a key event in the genesis of the lesions of atherosclerosis. *Science* **180**: 1332–1339.
9. Singer AJ, Clark RA. (1999) Cutaneous wound healing. *N Engl J Med* **341**: 738–746.
10. Hart J. (2002) Inflammation. 2: its role in the healing of chronic wounds. *J Wound Care* **11**: 245–249.
11. Pasterkamp G, de Kleijn DP, Borst C. (2000) Arterial remodeling in atherosclerosis, restenosis and after alteration of blood flow: potential mechanisms and clinical implications. *Cardiovasc Res* **45**: 843–852.
12. Bucala R, Spiegel LA, Chesney J, *et al.* (1994) Circulating fibrocytes define a new leukocyte subpopulation that mediates tissue repair. *Mol Med* **1**: 71–81.
13. Grainger DJ. (2004) Transforming growth factor beta and atherosclerosis: so far, so good for the protective cytokine hypothesis. *Arterioscler Thromb Vasc Biol* **24**: 399–404.
14. Orlandi A, Bochaton-Piallat ML, Gabbiani G, Spagnoli LG. (2006) Aging, smooth muscle cells and vascular pathobiology: implications for atherosclerosis. *Atherosclerosis* (in press).
15. Orlandi A, Marcellini M, Spagnoli LG. (2000) Aging influences development and progression of early aortic atherosclerotic lesions in cholesterol-fed rabbits. *Arterioscler Thromb Vasc Biol* **20**: 1123–1136.
16. Zaman AG, Helft G, Worthley SG, Badimon JJ. (2000) The role of plaque rupture and thrombosis in coronary artery disease. *Atherosclerosis* **149**: 251–266.
17. Gawaz M, Neumann FJ, Dickfeld T, *et al.* (1997) Vitronectin receptor (alpha(v)beta3) mediates platelet adhesion to the luminal aspect of endothelial cells: implications for reperfusion in acute myocardial infarction. *Circulation* **96**: 1809–1818.

18. Massberg S, Enders G, Leiderer R, *et al.* (1998) Platelet-endothelial cell interactions during ischemia/reperfusion: the role of P-selectin. *Blood* **92**: 507–515.
19. Massberg S, Brand K, Gruner S, *et al.* (2002) A critical role of platelet adhesion in the initiation of atherosclerotic lesion formation. *J Exp Med* **196**: 887–896.
20. Yla-Herttuala S, Lipton BA, Rosenfeld ME, *et al.* (1991) Expression of monocyte chemoattractant protein 1 in macrophage-rich areas of human and rabbit atherosclerotic lesions. *Proc Natl Acad Sci USA* **88**: 5252–5256.
21. Nelken NA, Coughlin SR, Gordon D, Wilcox JN. (1991) Monocyte chemoattractant protein-1 in human atheromatous plaques. *J Clin Invest* **88**: 1121–1127.
22. Han J, Hajjar DP, Febbraio M, Nicholson AC. (1997) Native and modified low density lipoproteins increase the functional expression of the macrophage class B scavenger receptor, CD36. *J Biol Chem* **272**: 21654–21659.
23. Hegyi L, Skepper JN, Cary NR, Mitchinson MJ. (1996) Foam cell apoptosis and the development of the lipid core of human atherosclerosis. *J Pathol* **180**: 423–429.
24. Raines EW, Ross R. (1993) Smooth muscle cells and the pathogenesis of the lesions of atherosclerosis. *Br Heart J* **69**: S30–S37.
25. Hart J. (2002) Inflammation. 1: its role in the healing of acute wounds. *J Wound Care* **11**: 205–209.
26. Leonarduzzi G, Sevanian A, Sottero B, *et al.* (2001) Up-regulation of the fibrogenic cytokine TGF-beta1 by oxysterols: a mechanistic link between cholesterol and atherosclerosis. *Faseb J* **15**: 1619–1621.
27. Grainger DJ, Kemp PR, Liu AC, *et al.* (1994) Activation of transforming growth factor-beta is inhibited in transgenic apolipoprotein(a) mice. *Nature* **370**: 460–462.
28. Ignotz RA, Massague J. (1986) Transforming growth factor-beta stimulates the expression of fibronectin and collagen and their incorporation into the extracellular matrix. *J Biol Chem* **261**: 4337–4345.
29. Simper D, Stalboerger PG, Panetta CJ, *et al.* (2002) Smooth muscle progenitor cells in human blood. *Circulation* **106**: 1199–1204.
30. Gordon D, Reidy MA, Benditt EP, Schwartz SM. (1990) Cell proliferation in human coronary arteries. *Proc Natl Acad Sci USA* **87**: 4600–4604.

31. Lutgens E, de Muinck ED, Kitslaar PJ, *et al.* (1999) Biphasic pattern of cell turnover characterizes the progression from fatty streaks to ruptured human atherosclerotic plaques. *Cardiovasc Res* **41**: 473–479.
32. Schwartz SM. (1999) The intima: a new soil. *Circ Res* **85**: 877–879.
33. Sata M. (2003) Circulating vascular progenitor cells contribute to vascular repair, remodeling, and lesion formation. *Trends Cardiovasc Med* **13**: 249–253.
34. Caplice NM, Bunch TJ, Stalboerger PG, *et al.* (2003) Smooth muscle cells in human coronary atherosclerosis can originate from cells administered at marrow transplantation. *Proc Natl Acad Sci USA* **100**: 4754–4759.
35. Sata M, Saiura A, Kunisato A, *et al.* (2002) Hematopoietic stem cells differentiate into vascular cells that participate in the pathogenesis of atherosclerosis. *Nat Med* **8**: 403–409.
36. Abe R, Donnelly SC, Peng T, *et al.* (2001) Peripheral blood fibrocytes: differentiation pathway and migration to wound sites. *J Immunol* **166**: 7556–7562.
37. Quan TE, Cowper S, Wu SP, *et al.* (2004) Circulating fibrocytes: collagen-secreting cells of the peripheral blood. *Int J Biochem Cell Biol* **36**: 598–606.
38. Mori L, Bellini A, Stacey MA, *et al.* (2005) Fibrocytes contribute to the myofibroblast population in wounded skin and originate from the bone marrow. *Exp Cell Res* **304**: 81–90.
39. Lutgens E, Gijbels M, Smook M, *et al.* (2002) Transforming growth factor-beta mediates balance between inflammation and fibrosis during plaque progression. *Arterioscler Thromb Vasc Biol* **22**: 975–982.
40. Kakolyris S, Karakitsos P, Tzardi M, Agapitos E. (1995) Immunohistochemical detection of fibronectin in early and advanced atherosclerosis. *In Vivo* **9**: 35–40.
41. Stenman S, von Smitten K, Vaheri A. (1980) Fibronectin and atherosclerosis. *Acta Med Scand Suppl* **642**: 165–170.
42. Moore BB, Kolodsick JE, Thannickal VJ, *et al.* (2005) CCR2-mediated recruitment of fibrocytes to the alveolar space after fibrotic injury. *Am J Pathol* **166**: 675–684.
43. Fletcher JP, Vasista A, Guiffre AK, Medbury HJ. (2004) Fibrocytes in atherosclerosis. *Circulation* **110**: 78.
44. Medbury HJ, Guiffre AK, Williams M, *et al.* (2006) The opposing roles of monocytes in atherosclerosis. *Submitted.*

45. Prockop DJ, Sieron AL, Li SW. (1998) Procollagen N-proteinase and procollagen C-proteinase. Two unusual metalloproteinases that are essential for procollagen processing probably have important roles in development and cell signaling. *Matrix Biol* **16**: 399–408.
46. Yang L, Scott PG, Dodd C, *et al.* (2005) Identification of fibrocytes in postburn hypertrophic scar. *Wound Repair Regen* **13**: 398–404.
47. McDonald JA, Kelley DG, Broekelmann TJ. (1982) Role of fibronectin in collagen deposition: Fab′ to the gelatin-binding domain of fibronectin inhibits both fibronectin and collagen organization in fibroblast extracellular matrix. *J Cell Biol* **92**: 485–492.
48. Schmidt M, Sun G, Stacey MA, *et al.* (2003) Identification of circulating fibrocytes as precursors of bronchial myofibroblasts in asthma. *J Immunol* **171**: 380–389.
49. Mach F, Schonbeck U, Bonnefoy JY, *et al.* (1997) Activation of monocyte/macrophage functions related to acute atheroma complication by ligation of CD40: induction of collagenase, stromelysin, and tissue factor. *Circulation* **96**: 396–399.
50. Davies MJ. (2000) Pathophysiology of acute coronary syndromes. *Indian Heart J* **52**: 473–479.
51. Diegelmann RF, Evans MC. (2004) Wound healing: an overview of acute, fibrotic and delayed healing. *Front Biosci* **9**: 283–289.
52. Varcoe RL, Mikhail M, Guiffre AK, *et al.* (2006) The role of the fibrocyte in intimal hyperplasia. *J Thromb Haemost* **4**: 1125–1133.
53. Blakytny R, Ludlow A, Martin GE, *et al.* (2004) Latent TGF-beta1 activation by platelets. *J Cell Physiol* **199**: 67–76.
54. Burke AP, Kolodgie FD, Farb A, *et al.* (2001) Healed plaque ruptures and sudden coronary death: evidence that subclinical rupture has a role in plaque progression. *Circulation* **103**: 934–940.
55. Braganza DM, Bennett MR. (2001) New insights into atherosclerotic plaque rupture. *Postgrad Med J* **77**: 94–98.
56. Khalil N. (1999) TGF-beta: from latent to active. *Microbes Infect* **1**: 1255–1263.
57. Grainger DJ, Kemp, PR, Metcalfe JC. (1995) The serum concentration of active transforming growth factor-beta is severely depressed in advanced atherosclerosis. *Nat Med* **1**: 74–79.
58. Grainger DJ, Byrne CD, Witchell CM, Metcalfe JC. (1997) Transforming growth factor beta is sequestered into an inactive pool by lipoproteins. *J Lipid Res* **38**: 2344–2352.

59. Martin-Paredero V, Vadillo J, Diaz J, *et al.* (1998) Fibrinogen and fibrinolysis in blood and in the arterial wall: its role in advanced atherosclerotic disease. *Cardiovasc Surg* **6**: 457–462.
60. Mallat Z, Gojova A, Marchiol-Fournigault C, *et al.* (2001) Inhibition of transforming growth factor-beta signaling accelerates atherosclerosis and induces an unstable plaque phenotype in mice. *Circ Res* **89**: 930–934.
61. Gamble JR, Khew-Goodall Y, Vadas MA. (1993) Transforming growth factor-beta inhibits E-selectin expression on human endothelial cells. *J Immunol* **150**: 4494–4503.
62. Bogdan C, Nathan C. (1993) Modulation of macrophage function by transforming growth factor beta, interleukin-4, and interleukin-10. *Ann N Y Acad Sci* **685**: 713–739.
63. Draude G, Lorenz RL. (2000) TGF-beta1 downregulates CD36 and scavenger receptor A but upregulates LOX-1 in human macrophages. *Am J Physiol Heart Circ Physiol* **278**: H1042–H1048.
64. Gojova A, Brun V, Esposito B, *et al.* (2003) Specific abrogation of transforming growth factor-beta signaling in T cells alters atherosclerotic lesion size and composition in mice. *Blood* **102**: 4052–4058.
65. Robertson AK, Rudling M, Zhou X, *et al.* (2003) Disruption of TGF-beta signaling in T cells accelerates atherosclerosis. *J Clin Invest* **112**: 1342–1350.
66. Hall MC, Young DA, Waters JG, *et al.* (2003) The comparative role of activator protein 1 and Smad factors in the regulation of Timp-1 and MMP-1 gene expression by transforming growth factor-beta 1. *J Biol Chem* **278**: 10304–10313.
67. Strassmann G, Cone JL, Arthur PM, *et al.* (1989) Effect of platelet-derived transforming growth factor (TGF) type beta 1 on murine inflammatory mononuclear phagocytes: increased fibronectin production. *Cell Immunol* **121**: 306–316.
68. Baccante G, Mincione G, Di Marcantonio MC, *et al.* (2004) Pravastatin up-regulates transforming growth factor-beta1 in THP-1 human macrophages: effect on scavenger receptor class A expression. *Biochem Biophys Res Commun* **314**: 704–710.
69. Falk E, Shah PK, Fuster V. (1995) Coronary plaque disruption. *Circulation* **92**: 657–671.
70. Davies MJ, Richardson PD, Woolf N, *et al.* (1993) Risk of thrombosis in human atherosclerotic plaques: role of extracellular lipid, macrophage, and smooth muscle cell content. *Br Heart J* **69**: 377–381.

71. Felton CV, Crook D, Davies MJ, Oliver MF. (1997) Relation of plaque lipid composition and morphology to the stability of human aortic plaques. *Arterioscler Thromb Vasc Biol* **17**: 1337–1345.
72. Pilling D, Buckley CD, Salmon M, Gomer RH. (2003) Inhibition of fibrocyte differentiation by serum amyloid P. *J Immunol* **171**: 5537–5546.
73. Kuwana M, Okazaki Y, Kodama H, *et al.* (2003) Human circulating $CD14^+$ monocytes as a source of progenitors that exhibit mesenchymal cell differentiation. *J Leukoc Biol* **74**: 833–845.
74. Zhao Y, Glesne D, Huberman E. (2003) A human peripheral blood monocyte-derived subset acts as pluripotent stem cells. *Proc Natl Acad Sci USA* **100**: 2426–2431.
75. Huzoor A, Ardlie NG. (1977) Platelet activation in haemostasis: role of thrombin and other clotting factors in platelet-collagen interaction. *Haemostasis* **6**: 59–71.
76. Davies MG, Hagen PO. (1994) Pathobiology of intimal hyperplasia. *Br J Surg* **81**: 1254–1269.
77. Shi Y, O'Brien JE, Fard A, *et al.* (1996) Adventitial myofibroblasts contribute to neointimal formation in injured porcine coronary arteries. *Circulation* **94**: 1655–1664.
78. Campbell JH, Han, CL, Campbell GR. (2001) Neointimal formation by circulating bone marrow cells. *Ann N Y Acad Sci* **947**: 18–24; discussion 24–25.
79. Han CI, Campbell GR, Campbell JH. (2001) Circulating bone marrow cells can contribute to neointimal formation. *J Vasc Res* **38**: 113–119.

Chapter 11

Nephrogenic Systemic Fibrosis: A Prototype Fibrocyte Disease

Cynthia L. Kucher and Shawn E. Cowper*

Introduction

In 1997, the first cases of an unusual new disease emerged. The condition affected only people with significant renal insufficiency and was characterized by thickening and hardening of the skin of the extremities. In some patients the disorder was rapidly progressive, resulting in a marked loss of mobility. Many of these patients became wheelchair-dependent in as little as a few weeks. Biopsies of affected skin showed significant fibrosis, with increased dermal fibroblast-like cells and markedly increased collagen, elastin and mucin production. With little insight into the cause of this disorder, it was given the descriptive name: "nephrogenic fibrosing dermopathy" (NFD). Now, nearly a decade later, further investigation and reasoned speculation have yielded evidence to support the notion that NFD is mediated by circulating fibrocytes (CFs), and is therefore a systemic condition. This crucial recognition has ramifications not

*Corresponding author: Yale Dermatopathology Service, 15 York Street, LMP 5031, PO Box 208059, New Haven, CT 06520–8059. Tel.: 203-785-4094; Fax: 203-785-6869. E-mail: shawn.cowper@yale.edu

only for NFD (and its future treatment), but possibly for the treatment and understanding of allied fibrosing disorders in other organ systems. Because of the recently-recognized systemic nature of the condition, the name "nephrogenic systemic fibrosis" (NSF) has been adopted.

Historical Context

In 1997, nephrologists at a Southern California renal transplant center began encountering patients with extensive cutaneous induration. All of these patients had in common a recent transplant procedure, and many of them had subsequently rejected their graft.[1] Histopathologic evaluation of the indurated areas revealed dermal and subcutaneous fibrosis associated with abundant mucin deposition. The histologic picture most resembled scleromyxedema (SCX); however, the clinical distribution of lesions (favoring the extremities and sparing the face) was the opposite of that expected in SCX. In addition, the absence of a circulating paraprotein (a finding present in most cases of scleromyxedema) further argued against SCX as an explanation.

The recognition of additional cases at other renal transplant centers prompted a letter to the Lancet in 2000,[2] and one year later, the publication of a thorough discussion of the histologic and clinical features of this new entity in a cohort of 14 patients.[3] Evaluation of this larger patient group revealed that while all patients with NSF had renal disease, neither transplantation nor dialysis were prerequisites for developing the disorder. This was contrary to the prevailing suspicion by some investigators (and patients) that dialysis was causing the disorder.[3–6] In fact, approximately 10% of patients who develop NSF have *never* been dialyzed.[3] A broader search for a common denominator resulted in the recognition that *without exception* all NSF patients have renal dysfunction.[7]

The California Department of Health Services, in cooperation with the Centers for Disease Control and Prevention conducted a case-control study among eight NSF patients, all of whom had renal disease and had undergone renal transplantation.[8] Compared with controls, the case-patients were more likely to have poor renal function post-transplantation, which included requiring hemodialysis and receiving medications associated with a severe disease. Of importance, no medication or infectious agent was identified as being causative (within the limitations imposed by the small scale of the study).

The ongoing investigation of NSF hinged on developing a larger epidemiological study. The NSF Registry project at Yale University (a direct extension of the original epidemiological studies) has confirmed over 190 cases from 29 states in the US[9] and a variety of European countries. Numerous additional reports have appeared in the medical literature, some from Europe and Asia.[10–13]

The Affected Population

NSF is equally represented among males and females with renal disease. It can affect persons of any age, including children and the elderly, but tends to affect the middle-aged most commonly. Many races have been represented among the stricken, including Caucasian, Asian, African, Indian, Pacific Islanders and Hispanic. Most reports have come from the United States, with a minority of reports from Europe, the Middle East, and Asia. Of interest, no confirmed Canadian or Australian cases have been published or entered into the Registry.[9–11,14–16]

Renal Disease

As mentioned above, the only immutable epidemiological feature of NSF is renal insufficiency.[3–5,8,9,16–19] While the renal disease may be acute, chronic or transient, the vast majority of patients suffer from chronic renal insufficiency and are usually dialysis-dependent. In all cases evaluated so far, the renal disease predates or arises concurrently with the skin lesions of NSF. Neither the cause, duration, nor severity of the underlying renal disease is directly related to the severity of the NSF.

One commonly encountered clinical scenario is the emergence of NSF during an acute worsening of an otherwise stable chronic renal

Table 1 Reported Causes of Renal Disease in NSF[14]

Acute interstitial nephritis
Atherosclerosis/Atheroembolism
Bright's syndrome
Chronic reflux/Ureteral valves
Cyclosporine toxicity
Diabetes mellitus
Familial nephrolithiasis
Goodpasture's syndrome
HELLP syndrome
Hepatorenal syndrome
Hypertension
Idiopathic
Polycystic disease
Postpartum eclampsia
Post-streptococcal glomerulonephritis
Primary hyperoxaluria I
Renal artery thrombi/emboli
Renal cell carcinoma
Scleroderma
Systemic lupus erythematosus
Takayasu's arteritis
Thrombotic thrombocytopenic purpura
Von Hippel Lindau syndrome
Wegener's granulomatosis

disease. When this occurs, disease onset is often tied to other clinical triggering events (see DISCUSSION), and is sometimes associated with new onset blood pressure lability.[16,20]

Specific causes of renal disease in patients with NSF are widely variable (Table 1).[3–5,9,14,17,19,21,22] In cases associated with reversible renal dysfunction (including successful renal transplantation), the return of normal renal function usually heralds an improvement in the cutaneous findings.

Dialysis

Approximately 90% of patients who develop NSF are dialysis patients.[9] Many of these patients show the first signs of NSF just as their renal disease reaches sufficient severity to warrant dialysis, sometimes creating the clinical impression that dialysis may be causing the NSF. Arguments against a cause and effect relationship with dialysis include the following facts: (1) 10% of NSF patients have never been dialyzed; (2) Many patients with NSF develop the disease after years to decades of uncomplicated dialysis treatments; (3) Patients developing NSF may be receiving dialysis in a variety of ways (i.e. inpatient or outpatient hemodialysis or peritoneal dialysis); (4) No single filter, manufacturer, dialysis unit, technique or dialysate has been consistently associated with NSF; (5) Many thousands of renal patients receiving dialysis never develop NSF.[3,18–20]

Reassuring patients that the life-sustaining technique of dialysis does not promote the development of NSF, and that discontinuance of dialysis will not result in a cessation of NSF, is of paramount importance.[7] Unfortunately, establishment of a stable dialysis regimen does not appear to impede the progression of NSF.

Renal transplantation

If renal function can be reestablished, either through renal transplantation or by supportive treatment of reversible renal processes, an improvement in the clinical findings of NSF is typical.[7] Lack of full reversibility may be seen in longstanding cases of NSF (typically over one year) and in patients in whom extremity contractures have become well-established. For this reason, expeditious reestablishment of renal function, when possible, is strongly recommended. For some patients this may require an accelerated placement of a renal graft as well as aggressive supporting physical therapy until the new kidney can be transplanted.[7,20] NSF will sometimes develop shortly after renal transplantation. The likely reasons for this are discussed below (see DISCUSSION). In general, if the transplanted kidney is healthy and not rejected, the disease will improve without further NSF-directed therapy.

Other Comorbidities

Hypercoagulability, thrombosis, and endothelial injury

A growing number of NSF patients are recognized as having hypercoagulability and/or thrombotic events that temporally relate to the onset of the cutaneous disease. In some cases, a clinically unsuspected hypercoagulable state is first recognized after the onset of NSF.[14] Positive anti-cardiolipin antibodies have been identified in some NSF patients, but are not universal.[4,9] Elevated levels of these antibodies are known to result in hypercoagulable states. However, increased titers of these antibodies are regularly detected in many patients with end-stage renal disease (10–29%), making it difficult to know how or whether they contribute to the development of NSF. It has been suggested that the lipid molecule in these antibodies somehow interacts with the lipid substance in the tubing of the dialysis equipment to stimulate fibroblast or mucin production. Of course, this explanation would not account for the subset of patients who develop NSF without any prior dialysis exposure, or those who do not harbor these antibodies.

Upon careful questioning and chart review, most patients are found to have sustained a significant thrombotic event in the two weeks prior to NSF onset. The causes for the thrombosis are widely variable (Table 2)[14] and may or may not be related to a measurable (or clinically recognized) hypercoagulable state. Of those who have undergone renal transplantation, many sustain thrombotic loss of the newly transplanted kidney, followed by the development of NSF. Others develop NSF concurrently with the progression of lower extremity deep venous thrombosis that extends proximally to involve the inferior vena cava and/or renal veins.

Many NSF patients also report non-transplant related surgical procedures immediately before disease onset. Within this group, the surgical procedure often has an associated vascular reconstruction component. If transplantation procedures (hepatic or renal) are included in the cohort, the number of NSF patients who have had surgery within two weeks of the disease onset increases to 48%. If

Table 2 Reported Thrombotic Events and Hypercoagulable States

Thrombosed arteriovenous fistula
Deep venous thrombosis
Pulmonary embolus
Atrial thrombus
Femoral graft thrombosis
Factor V Leiden
Antithrombin III deficiency
Anticardiolipin antibody
Protein S deficiency
Protein C deficiency
Hyperhomocysteinemia

dialysis fistula construction or central catheter placement is included as well, the percentage jumps to 90%.[14] A significant number of patients describe clotting of an arteriovenous fistula shortly before developing NSF. Others report thrombosis of indwelling ports or PICC (peripherally inserted central catheter) lines that seem to serve as an epicenter for NSF that subsequently becomes more generalized in its distribution.[16]

While the exact significance of the above associations is unclear, it has become increasingly apparent that vascular injury, either as a result of manipulation or thrombosis (or both), likely sets in motion a chain of inflammatory signals that replicate the early stages of wound healing, eventuating in the recruitment of the cells responsible for the aberrant peripheral collagen and matrix deposition seen in NSF.[16,20] If borne out, this conclusion suggests that anticoagulation therapy (possibly administered prophylactically) may prove beneficial in the treatment or prevention of NSF.

In the event that a patient with known NSF is anticipating renal transplantation, a careful hypercoagulation workup should precede surgery, and appropriate precautions should be taken to avoid thrombotic loss of the new renal graft.[7]

Other systemic processes

Many patients with NSF have chronic hepatic disease.[4,5,14,19,21,22] The diseases most commonly encountered are hepatitis B- or C-induced cirrhosis. In some clinical situations, the hepatic disease induces hepatorenal syndrome which then serves as the underlying cause of renal dysfunction. If these patients also develop NSF, reestablishment of renal function via a successful hepatic transplantation may result in an improvement in the cutaneous lesions.[9,16,19]

Underlying chronic pulmonary fibrosis, often of an idiopathic nature, is sometimes seen in NSF patients. In the late stages of severe NSF involving the chest wall, a restrictive pulmonary defect may develop. It is not clear whether this is due to chest wall thickening, or diaphragmatic and/or pulmonary parenchymal fibrosis.[14,16,20,23] The pulmonary dysfunction seen in NSF is more insidious than the directly observable cutaneous lesions, and ascribing the same cause and effect relationship with renal dysfunction is merely implied, not firmly established.[24]

Among patients who have supplied detailed histories, two have known brain tumors. As brain tumors in the general population are rare, this finding appears unusual, and is therefore mentioned here.[14]

Autopsy studies have revealed the presence of widespread cardiac fibrosis in two young patients with NSF.[9,25] Cardiac fibrosis is a common autopsy finding in patients with hypertension and diabetes (frequently comorbid conditions with renal disease), but the presence of such widespread disease in these young NSF patients is distinctly unusual, and judged to be due to the effects of NSF. The degree of cardiac dysfunction due to fibrosis may be clinically unsuspected, as it may not be externally measurable. In addition, fibrosis has been described in the dura mater, proximal esophagus, psoas muscles, and contiguous fascia and skeletal muscles of the overlying dermal plaques.[16,26,27]

An element of caution is necessary in the interpretation of fibrosis in NSF patients, however. To claim confidently that the finding of fibrosis found on autopsy examination of an affected patient is due to NSF is, at present, somewhat of a logical leap. The fibrosis found in these patients must be considered in the context of the patient's prior medical history, the onset and course of the NSF, and the degree and temporal course of medical dysfunction of that organ. We believe the evidence suggests that fibrosis of other organ systems in NSF is at least partly due to the same processes taking place in the skin for the following reasons: (1) as noted above, pulmonary fibrosis appears to be present in a disproportionate number of patients with NSF; (2) young patients without other known cardiac processes may exhibit extreme cardiac fibrosis; (3) radiologic and nuclear medicine examinations suggest abnormal findings in soft tissues throughout the body (see ancillary studies, below); and (4) NSF is mediated by a *circulating* cell capable of initiating the changes observed histologically.[18,28]

Presentation

Signs, Symptoms and Progression

The lesions of NSF are typically symmetrical, and develop on the limbs and the trunk. A common distribution is between the ankles and mid-thighs, and between the wrists and mid-upper arms, bilaterally. Occasionally, swelling of the hands and feet, sometimes associated with bullae, is noted. The primary lesions are skin-colored to erythematous papules that coalesce into erythematous to brawny plaques with a "peau d'orange" appearance.[3,14,20] These plaques have been described as having an "amoeboid" advancing edge (Fig. 1).[4] Nodules are sometimes also described.

The involved skin becomes markedly thickened with a woody texture. The tautness of the skin sometimes imparts a shiny surface, especially in areas where hair loss occurs. The time frame in which the condition develops can vary widely, with some ambulating patients becoming wheelchair dependent within days to weeks of onset. Plantar flexion of the feet may be severe enough to make ambulation impossible.[3,14,20] Often, the changes are so severe as to inhibit the flexion and extension of joints, resulting in contractures (Fig. 2).

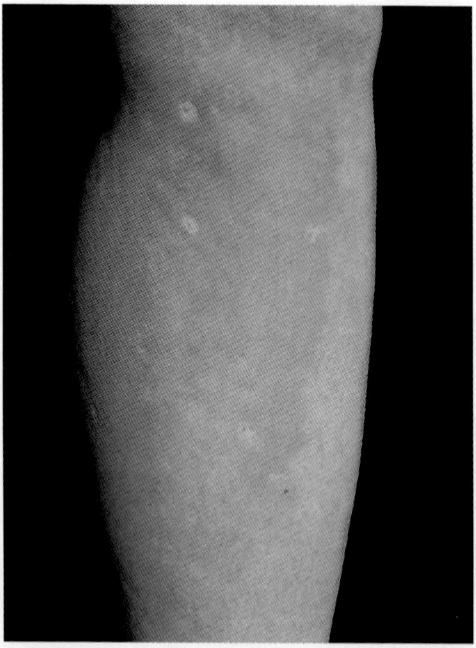

Fig. 1. Lower extremity demonstrating erythematous to brawny plaque with a "peau d'orange" texture.

Patients may complain of pruritus and sharp pain in the affected areas. Others describe sharp pain in the rib cage simulating costochondritis. Arthritis, fever and myalgias are not seen, although muscle weakness has been described. Because of the acute presentation and appearance of early lesions, it is not uncommon for patients to be mistakenly treated for cellulitis. MacKay-Wiggan, *et al.* reported a common scenario of marked extremity swelling that, once resolved, reveals the characteristic woody consistency described above.[4] Occasional patients (estimated at less than 5%) have rapidly progressive, fulminant NSF associated with an accelerated loss of mobility, and often severe pain. Some patients describe wildly fluctuating blood pressure about the time of disease onset.[20]

Streams *et al.* have described yellow scleral plaques in two patients with NSF.[17] Although not universal, these plaques are commonly present. They do not seem to affect the vision, and are

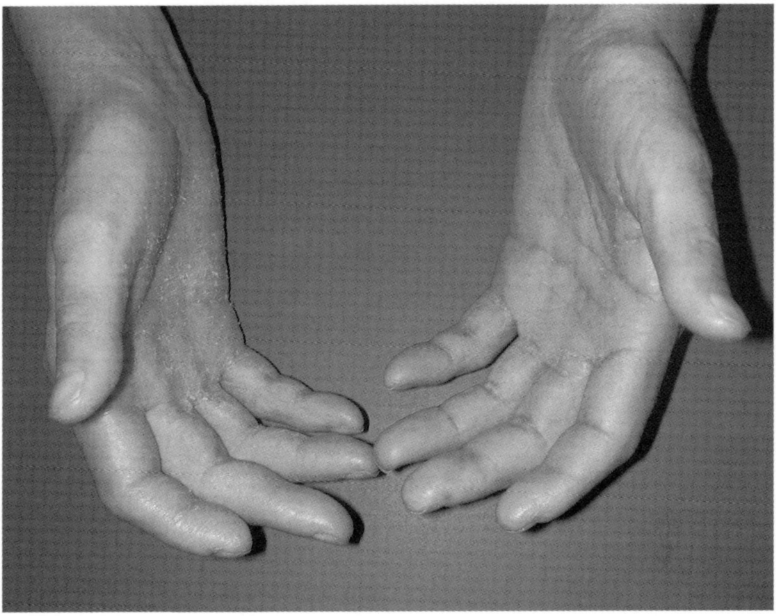

Fig. 2. Flexural deformity in the hands of a patient with nephrogenic systemic fibrosis.

not painful (Fig. 3). What relationship they have to NSF is unclear at present, although observers agree they are not specific to the disease process. The histopathologic findings of these plaques have yet to be described.[14,16,20]

Diagnosis

There is no single diagnostic test used for NSF. When a patient is observed with the above noted constellation of findings in the setting of renal insufficiency, a confident diagnosis can usually be reached.[14,20] Diagnoses that often enter the clinical differential are listed in Table 3.[3,14,16]

Laboratory investigation

The laboratory data in this patient population show abnormalities of renal function. Patients with renal insufficiency have numerous

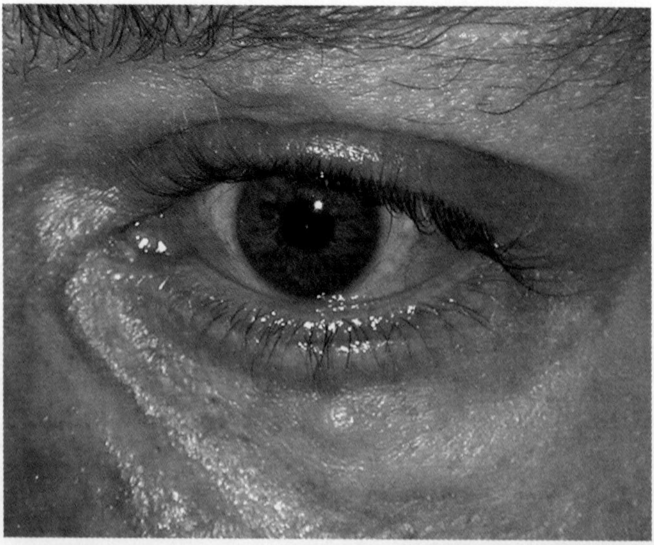

Fig. 3. Conjunctival injection and white-yellow scleral plaques in a patient with nephrogenic systemic fibrosis.

Table 3 Clinical Differential Diagnoses for NSF

Scleromyxedema
Eosinophilic fasciitis
Eosinophilia-myalgia syndrome
Toxic oil syndrome
Sclerodermoid graft versus host disease
Fibrosis (induced by drugs, silica or organic solvents)
Fibroblastic rheumatism
Borreliosis
β_2-microglobulin amyloidosis
Systemic sclerosis/morphea
Scleredema of Buschke
Amyloidosis
Carcinoid syndrome
Porphyria cutanea tarda
Calciphylaxis
Lipodermatosclerosis

predictable laboratory abnormalities based upon the degree of renal disease and the nature of comorbid conditions.[14] In addition, as noted above, hypercoagulable states (often clinically unsuspected) are common, and wide fluctuations in platelet counts have been observed.[9,14] Antinuclear antibodies and other serologic tests are usually not observed (unless the patient has renal disease due to autoimmune conditions such as systemic lupus erythematosus).[14,20] Several patients with very high (and unexplained) alkaline phosphatase levels have been noted.[16] The cause of this has not been determined. Suggested laboratory tests that may be helpful in distinguishing NSF from clinically similar conditions are summarized in Table 4.[16]

Biopsy and histopathology

The histopathological examination of a skin biopsy specimen from lesional skin is the gold standard for diagnosis of NSF. An incisional or deep punch biopsy which contains representative epidermis, dermis, subcutaneous fat, and possibly fascia and skeletal muscle is necessary for a definitive diagnosis.[14]

Table 4 Laboratory Tests of Value in the Workup of NSF

Absolute eosinophil count
Platelet count
Coagulopathy studies (PT/PTT, Protein C and S, Antithrombin III, homocysteine, factor V Leiden)
BUN and creatinine
Liver panel
Serum protein electrophoresis and immunoelectropheresis
Urine electrophoresis
Thyroid function tests
Erythrocyte sedimentation rate
Antinuclear antibody (IgG subclass)
Serological autoantibody panels (to include Scl-70, anti-centromere, and anti-RNP)
Antiphospholipid antibody (Anticardiolipin Antibody)
Hepatitis B and C serologies
Urine porphyrin studies

Biopsies obtained from lesional skin commonly reveal numerous CD34/procollagen-I dual positive dermal spindle cells (fibrocytes) diffusely arrayed within the mid-to deep dermis, extending into widened interlobular septa of the subcutaneous fat. (Fig. 4A-B) On very deep incisional biopsies, these cells may extend into the fascia and underlying skeletal muscle.

The fibrocytes appear as elongated spindle cells with tapering nuclei that form a dense interconnecting network, entwining elastic fibers and collagen bands. CD34 staining is membranous, whereas that of procollagen I is cytoplasmic (Fig. 4C-D). Reticular dermal collagen bundles are often thickened (and occasionally hyalinized) and

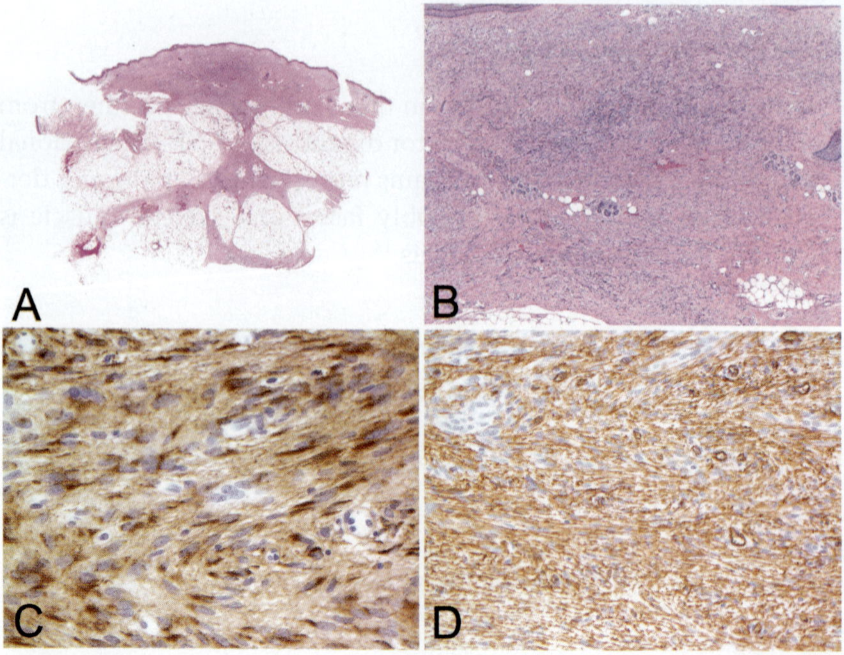

Fig. 4. **(A)** Scanning magnification of deep incisional skin biopsy showing marked dermal infiltration by spindle cells, increased collagen deposition, and widening of subcutaneous septa (H&E stain). **(B)** Skin biopsy showing dense infiltration by spindle and epithelioid cells (H&E stain, 100x). **(C)** Cytoplasmic staining with immunohistochemical stain directed against procollagen I (400x). **(D)** Membranous staining of same spindle cells with CD34 (200x).

separated from one another by prominent clefts. The large bundles are commonly in contact with, and often contain, the dendritic processes of fibrocytes. These cells may also be identified within small clefts inside the thick bundles. CD34 staining also reveals that many of these large collagen bands are subdivided into smaller polygonal fibers by fine dendritic processes, conferring an overall architecture reminiscent of skeletal muscle fascicles.[3]

The background elastic fibers are commonly prominent and thick, and may be appreciated on H&E sections as refractile, glassy eosinophilic structures. These fibers tend to orient adjacent to the fascicular collagen, forming layers that are roughly parallel to the epidermis. In addition, very fine elastic fibers may be found focally within the fascicles.

Often, numerous epithelioid to stellate factor XIIIa positive cells, sometimes in the company of CD68 positive mono- and multinucleated cells (Fig. 5), form small clusters. On occasion, these are the dominant cell type. These foci often contain abundant mucin as well, and can be found within the widened septa of fatty lobules. Occasionally hemosiderin and calcium are noted near these foci, and in rare cases, ossification may be noted.[13] Sometimes factor XIIIa^{+} multinucleated cells (with 10–12 nuclei) are seen adjacent to CD68^{+} multinucleated cells (with 3–4 nuclei). In occasional cases, factor XIIIa positive epithelioid and stellate cells may dominate the picture.

CD34 staining occasionally reveals inconspicuous angiogenesis within fatty septa. In addition, adipocytes adjacent to the septa may be atrophic, and rarely, giant cells containing clefts resembling Miescher's radial granulomas may be seen. A very sparse superficial and deep perivascular lymphocytic infiltrate extending as deep as the fatty lobules is sometimes noted. Plasma cells are typically absent, and only rare cases contain eosinophils. Appendageal structures may be reduced in number.

Swartz *et al.* have reported a population of myofibroblasts (smooth muscle actin positive) in NSF lesions between three- and four-weeks old. This population was absent in lesions over seven weeks old.[5,14]

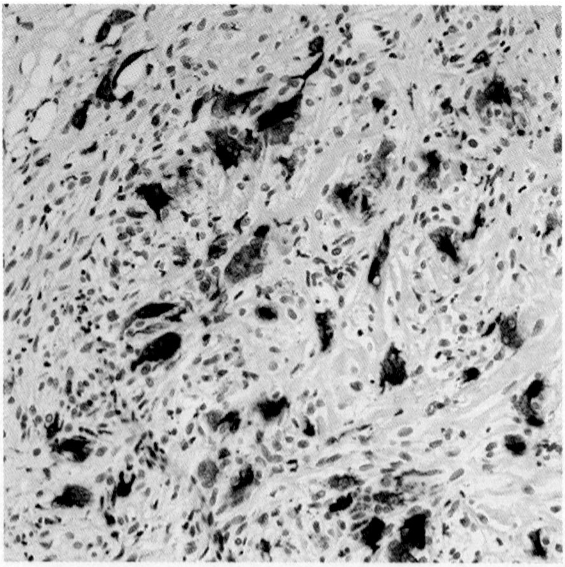

Fig. 5. Cluster of CD68 positive multinucleated histiocytes within an expanded subcutaneous septum (200x).

Ultrastructural examination has revealed numerous oval and spindle-shaped cells with long thin cytoplasmic processes (dendrites) with prominent rough endoplasmic reticulum (Fig. 6). Dendritic processes may be found within the collagen bundles. Some elastic fibers may be coated by an electron-dense material which may represent calcification.[3]

The main histopathologic differential diagnosis is with scleromyxedema,[29] although other differentials exist depending on the scope of the clinical history received, the adequacy of the biopsy, and the developmental stage of the disorder (Table 5).

Ancillary studies

Radiologic studies have been utilitzed to a limited extent in NSF. In a paper reported from Belgium,[30] magnetic resonance imaging (MRI) showed an "inhomogeneous" pattern of the sort commonly encountered in myositis. Ultrasound imaging of this patient showed diffuse

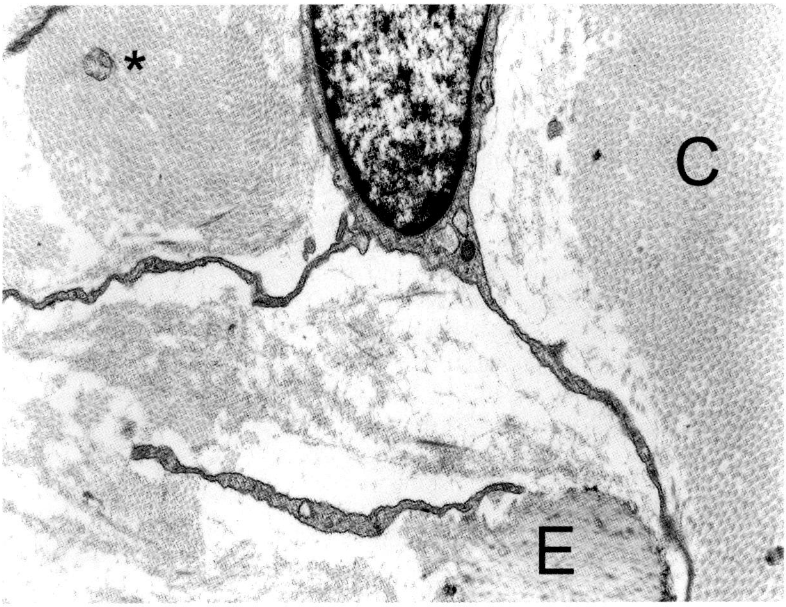

Fig. 6. Electron micrograph revealing central spindle cell nucleus with attached dendritic processes. The processes surround collagen bundles (C) and focally contact elastic fibers (E). One dendritic process in cross section (*) is present within a collagen bundle.

infiltration of the subcutaneous tissues. Because of the possible exacerbating effects of contrast agents utilized in MRI studies, however (see discussion), the use of this method of imaging should probably be limited to the greatest extent possible.

Two patients had whole body positron emission scanning (PET) which confirmed increased metabolic activity in all the clinically involved tissues as measured with the radioactive tracer 18F-fluorodeoxyglucose. These authors also noted abnormal uptake of the bone scintigraphy tracer Tc99m-MDP in involved soft tissues, a finding attributed to inflammation-induced loss of cell membrane activity.[16]

Of interest, two additional patients with contractures following hepatic transplantation for hepatorenal syndrome[31] showed increased uptake of another bone scintigraphy tracer, Tc99m-HDP.

Table 5 Histopathological Differential Diagnoses for NSF

Scleromyxedema
Eosinophilic fasciitis
Eosinophilia-myalgia syndrome
Toxic oil syndrome
Sclerodermoid graft versus host disease
Fibroblastic rheumatism
β_2-microglobulin amyloidosis
Systemic sclerosis/morphea
Scleredema of Buschke
Amyloidosis
Carcinoid syndrome
Porphyria cutanea tarda
Calciphylaxis
Lipodermatosclerosis
Dermatofibrosarcoma protuberans
Melanoma (spindle cell variant)
Granuloma annulare

While the authors indicate both patients had a similar diffuse and symmetrical uptake of tracer, they diagnosed one of the patients with nephrogenic fibrosing dermopathy, and the other with graft versus host disease. The second patient never had a skin biopsy, which (we believe) would also have shown NSF.[16]

Several additional patients in another cohort[32] were scanned by CT and MRI techniques. The most severely affected patients showed generalized muscle atrophy and the presence of diffuse high attenuation material throughout the muscles and fascial planes of the entire body. This material was felt to be consistent with fibrosis.

Multiple patients in the above study also received electromyography (EMG) and nerve conduction velocity (NCV) testing. Most of the patients tested exhibited a sensory-motor polyneuropathy that rated as mild to moderate in severity. All of the patients manifested varying degrees of myopathy, and one of the patients also showed evidence of denervation injury.[16]

Treatment

Renal transplantation

Numerous therapies for NSF have been investigated, most of which have not shown significant efficacy. Improvement of renal function alone, regardless of modality, seems to slow the progression of the disease. In many cases, improvement of renal function has terminated the progression of NSF, and occasionally caused the disease to slowly resolve.[9] Given this observation, the best currently available option for treatment for NSF patients is renal transplantation. Many of these patients are already on transplant waiting lists, but with the development of NSF, strong consideration should be given to expediting this process to the extent possible before the NSF becomes sufficiently established to lead to permanent and irreversible disability. One important caveat is that a full hypercoagulability work-up should be performed prior to the transplant procedure so that clotting tendencies can be appropriately managed to avoid thrombotic loss of the new graft. The rationale for transplantation is discussed in greater detail elsewhere.[7]

Extracorporeal photopheresis (ECP)

Unfortunately, many NSF patients are not transplant candidates. For these patients, the best documented therapy to date is extracorporeal photopheresis (ECP). ECP has proven beneficial in three patients in Europe and limited success has been observed in the United States as well.[10,31,33] The patients in at least one of the European studies received treatment within one year of developing NSF symptoms. This is an important observation as ECP may have minimal to partial success in longstanding (> 1 year) disease.[9] The authors have seen improvement in one patient treated with this modality, and are aware of several others who have shown marked improvement (unpublished). Another patient with extensive longstanding disease failed to improve. Continued experience with ECP will certainly clarify the best clinical predictors of success with this modality. Currently the greatest barrier to treatment with ECP is its exorbitant cost. At present, only a few insurers cover this therapy; however, an appeal to the insurance company with reference to the developing literature on the subject will sometimes result in funding for a trial of therapy. Of interest, those who do respond generally show measurable improvement within a month or two of initiating the therapy. The low risk of ECP, teamed with the very striking improvements in quality of life a patient may experience, makes a trial of therapy a worthwhile venture.[20]

Plasmapheresis

In 2003, Baron *et al.* decribed several NSF patients who showed improvement following treatment with plasmapheresis.[19] Two of these patients exhibited a concurrent improvement in renal function (a complicating factor in terms of interpretation of the results). Other investigators found plasmapheresis to be partially effective[4] or completely ineffective.[22] Unfortunately, the degree of the underlying renal function clinical involvement, and duration of disease were incompletely documented in some of these reports, making it difficult to understand the true impact of plasmapheresis as a potential treatment for NSF.

Other considerations

Other treatments which have been tried, and have shown variable effectiveness, have included oral and topical steroids, high dose intravenous immunoglobulin therapy, and pentoxifylline.[9,27] Ultraviolet light therapy has been helpful in one reported patient.[34] Physical therapy, in particular swimming, may be useful in retarding the progression of the disease.[9] Thalidomide was of interest early on; however, use as a first-line agent is not recommended due to the lack of published trials, and a reported onset of NSF in one patient taking thalidomide for other reasons (authors' observation). Additional anecdotal agents include selective histamine blockade, calcipotriene ointment, cyclophosphomide, cyclosporine, interferon-alpha, psoralen ultraviolet light, intravenous immunoglobulin and physical therapy. Formal reports on the effectiveness of these agents are not yet available.[14]

Leboit suggests that decreasing the dose of erythropoietin (epo) in patients with renal dysfunction might improve NSF in some patients since recombinant epo has potential fibrogenic properties.[1] Anecdotal evidence suggests that a reduction in dose of epo in at least one patient has resulted in some clinical improvement.[20] Given this preliminary observation, it is reasonable to consider a reduction of dose in patients receiving this therapy.

Experts agree that poor renal function is a prerequisite for developing NSF. Alleged therapeutic successes must therefore be reported in the context of the patient's renal disease. If the patient has been dialysis-dependent throughout the course of treatment with the allegedly beneficial therapy, any clinical improvements seen in the NSF may indeed be due to the applied treatment. If the renal status is not provided, or if the patient's renal function has improved, the contribution of the allegedly beneficial therapy cannot be separated from the positive contributions of the improved baseline renal function, and a confident conclusion about the therapy cannot be rendered.[14] To date, ECP is the only therapy that satisfies these conditions.

Discussion

Circulating Fibrocytes

In 2003, Cowper, *et al.* reported that most of the dermal spindle cells found in the biopsies of involved cutaneous tissue are immunohistochemically positive for CD34 and procollagen I.[18] This immunological profile, not previously described in the skin, is identical to that of the "circulating fibrocyte" (CF).[20] Besides dual positivity for collagen and CD34, several other clues support the hypothesis that CFs are involved in NSF. 1) the symmetry of lesions suggests a circulating (intrinsic) factor is involved; 2) the rapidity of development of the lesions (with concurrent absence of mitotic figures among the spindle cells) implicates the recruitment of a blood-borne cell; 3) the concurrence of histiocytes in varying proportions in the infiltrate suggests a common origin for these cells (CFs are known to differentiate from monocytic precursors under certain conditions);[35] 4) the presence of fibrosis in other non-contiguous organ systems supports a systemic disorder; 5) histologically, the cellular composition of an active case of NSF closely resembles a wound healing reaction (including angiogenesis)[36]; 6) the resolved lesions of NSF are histologically indistinguishable from those of a healed wound.[18]

A basic conceptual model and possible triggers

In the early phases of normal wound healing, platelets, endothelium, perivascular connective tissues, and tissue-based and circulating factors all contribute to a complex cascade of events that eventuates in the recruitment of spindle cells capable of matrix deposition. Circulating fibrocytes are known to represent at least part of this spindle cell population in normal individuals.[37]

In NSF, circulating fibrocytes arrive in cutaneous soft tissues and develop into a fibrosing reaction histologically indistinguishable from the proliferative stages of wound healing. Unlike normal wound healing, however, NSF fibrocytes engage in this activity without a clinically-evident wound. In the event renal function can be restored, these cells generally downregulate. If the renal function cannot be improved, and effective therapy is not available, these changes may reach a point of irreversibility (even in the event later successful restoration of kidney function can be achieved).

These observations illustrate several important concepts: (1) the fibrocytes of NSF are capable of ordered and purposeful behavior that appears to be driven by local or systemic cell signaling; (2) return of normal kidney function restores normal signaling in the early phases of the disease; (3) the later stages of the disease may be under different controls (or simply irreversible due to deposition of vast amounts of matrix material). Because of the early reversibility, we infer that CFs are *capable* of normal function, and that their atypical behavior is a response to the abnormal signaling milieu that exists in some patients with renal disease. This observation also suggests that a careful study of *normal* CFs will likely yield clues about the causes of NSF and may provide insights into possible treatments. In much the same way that disseminated intravascular coagulation represents an inappropriate activation of the normal process of coagulation, we feel that NSF likely represents an inappropriate activation of the first stages of wound healing.

As mentioned above, patients with NSF have a high likelihood of harboring a hypercoagulable tendency. The high incidence of vascular trauma and/or thrombosis in the days surrounding the disease onset, and the observed phenomenon of NSF arising adjacent to indwelling catheters and along the course of PICC lines, suggest that the interaction of the endothelium with the platelets is of key importance in the early stages of NSF (just as it is in the early stages of wound healing).

As others have observed, microthrombi are generally not encountered in histological sections from the skin of NSF patients. This has been our experience as well. It is our contention that endothelial injury, possibly induced by thrombosis or the indwelling catheters, or perhaps secondary to contact with other endogenous or exogenous substances, may be all that is required to initiate NSF. If indeed there are humoral signals elaborated by endothelial injury, they may circulate to *any* location with a patent vasculature, but would most likely exert their effects in the vicinity of the injury.

The lesions of NSF tend to form in dependent sites, and are often most severe in tissues infiltrated by marked edema (due to obstruction of vascular return or low oncotic pressure). It seems plausible that any circulating profibrotic signal(s), simply as a consequence of increased "dwell time" in low flow areas, may be more persistent in the extremities than in other body sites. In fact, local flow diminution may also facilitate the exit of CFs from the circulation into the tissue, and may in turn also account for the tendency of NSF lesions to develop in dependent sites.

The newest suspect: Endothelin-1

A profibrotic peptide of great interest in the vascular biology community is endothelin-1 (ET-1). First described in 1988 by Yanagisawa *et al.*,[38] ET-1 is produced by the endothelium (and other tissues, including monocytes/macrophages) and is known to be chemotactic for CFs. ET-1 has been implicated in the pathophysiology of fibrosis in numerous organ systems, including the skin, lung and heart.[33,39,40] ET-1 mediates the production of collagens I and III and fibronectin (itself chemotactic for fibroblasts[33]), an effect that is thought to chiefly occur through ETA, the chief ET-1 receptor on human dermal fibroblasts.[40] ETA is also responsible for mediating collagen I degradation. Of interest, after prolonged stimulation by ET-1, dermal fibroblasts begin expressing more ETB receptor (primarily an inducer of collagen I synthesis). This receptor subtype shift results in the maintenance of production of collagen I, and shifts away from collagen III and fibronectin production and collagen I degradation. Increased dermal fibroblast expression of ETB has been described in systemic sclerosis.[40]

The most potent physiologic factor in regulating ET-1 production seems to be blood flow. Increased blood flow, by activation of shear stress receptors in the endothelium, results in a decrease in production and release of ET-1.[33] In addition, ET-1 is elevated in patients with renal disease as compared to normal controls, and is known to become elevated in response to vascular injury, and in the setting of reduced extracellular pH.[39] ET-1 can be effectively cleared with dialysis; however, investigators believe that ET-1 release *induced by the dialysis procedure itself* might exceed the filtration benefits of the procedure, yielding a higher net concentration of the peptide.[41] Conflicting reports have yet to resolve the net effects of various dialysis procedures on the levels of ET-1.[34]

ET-1 is also the most potent vasoconstrictor yet identified in the human organism,[42] and exerts its effects in the setting of chronic renal patients as well as in healthy human controls.[34] Because of its local effects, plasma levels of ET-1 may not correlate with the intensity of the resultant vasoconstriction.[34,39]

Plasma ET-1 levels are also known to increase in coronary artery bypass procedures.[42] Comparisons between the heparin coatings of two manufacturers of coronary bypass circuits have shown differences in their abilities to quell the rise of bypass-induced ET-1 elevations.[42] This finding suggests that the manufacturing techniques of bioactive surfaces may have an impact on the production of ET-1 during their use.

Fibrosis via accretion — a proposal

While a formal study of ET-1 in the setting of NSF has yet to be conducted, there are a variety of compelling and mutually supporting arguments that strongly implicate its role in mediating the disease. 1) ET-1 is chemotactic for CFs and is capable of inducing all of the key histological findings associated with NSF; 2) ET-1 is elevated in the setting of renal disease, is exacerbated by low flow states, and can be rapidly cleared with restoration of normal renal function; 3) ET-1 expression is elevated in vascular bypass surgery, and its degree of elevation may be influenced by the heparin coatings on synthetic surfaces; 4) ET-1 production is enhanced by hypoxia and reduced by treatment with pentoxifylline and angiotensin converting enzyme inhibitors (described by other authors as being useful therapeutic measures in selected patients with NSF[39,43,44]); 5) the vasoactive capacity of ET-1 may explain the unusually labile blood pressures that have been encountered at the onset of NSF in some patients.[20]

These observations by themselves do not constitute a proof, but do illustrate a compelling hypothesis for the role of ET-1 in NSF. In addition, while ET-1 induced effects could be a common denominator for all patients with NSF, this hypothesis does *not* require that all patients arrive at the elevated ET-1 levels in precisely the same way. We have commented before that, besides renal disease, no single factor has been identified that explains all cases of NSF. Based on the above arguments, it may be that no *single* factor will *ever* be found. NSF may simply be the result of a collection of profibrogenic tendencies that, in a process of accretion, finally build to clinically significant, rapidly developing fibrosis.

Fibrosis via exogenous substances — an alternate hypothesis

In the past, we have suggested that a newly introduced material, possibly a contrast agent, medication, or other "allergen" could be depositing in the peripheral tissues and serving as a surrogate target for CFs that were mobilized in response to other inciting events (e.g. surgery or tissue injury).[24] A sequential combination of events (renal disease > allergen deposition > CF release and mistargeting) could also explain all of the unanswered questions to date, including the finding of elevated TGF-β (produced by CFs)[27]; the association with surgical procedures and thrombosis[14]; the resemblance to wound healing in early and resolved stages[3,14]; the recent emergence of NSF (new allergen or allergens); the absence of a genetic predisposition[14]; and the presence of fibrosis in other organ systems (likely additional sites of allergen deposition). The model would also explain why NSF does not affect all renal patients as it presupposes that a specified sequence of events has to occur, and that (as in allergic asthma) hypersensitivity to a peripherally deposited agent exists in the patient.[24]

Whichever of these mechanisms is at play, a recent publication may provide a valuable clue. In Austria, the exposure of nine patients with NSF to the gadolinium-containing MRI contrast agent, gadodiamide, has been described.[43] In this study cohort, the authors point out that those patients with renal disease who received gadodiamide and developed NSF were more likely to be acidotic, as compared with those who did not develop NSF. The proposal is of great interest, as many patients in the NSF Registry also had gadolinium-containing contrast studies (although the exact formulation of the contrast has yet to be investigated). In addition, the early work done by the Centers for Disease Control did not investigate contrast agents, but did conclude that those who developed NSF were generally more critically ill (and hence, would be more likely to be acidotic).[8]

Clearly, the confluence of several lines of investigation suggests that endothelial injury may be a triggering event in those prone to NSF. Perhaps in a significant number of these patients, externally administered contrast agents (gadolinium-containing or others) are either triggering or potentiating local and systemic fibrosis through the mechanisms hypothesized above. Contrast agents are an attractive suspect, as they are increasingly administered in the evaluation of renal disease, in particular since 1997, when the first cases of NSF were encountered. Gadolinium-containing agents have been touted as being relatively kidney-safe for renal imaging, and are commonly used in the transplant perioperative period, in the evaluation of thrombosis, in the planning stages of vascular surgery, and in the evaluation of brain neoplasia — all clinical situations that are known to occur during the period temporally preceding the onset of NSF.

Further epidemiological investigation with case-matched controls will be a critical step in strengthening or refuting the cause and effect implications in regards to contrast. While no animal model for NSF currently exists, the addition of an animal model to the disease investigation would be of inestimable value in answering this question, as well as studying the potential effects of a variety of possible therapies. Of great interest is the current availability of an ET-1 blockading agent, bosentan. While not formally tested in the NSF population, there is great interest in determining whether this non-selective agent (or other selective ETA and ETB blockading agents currently in clinical trials) might be useful in the treatment of NSF or other fibrosing disorders. These agents are also under intense investigation for the reduction of fibrosis-induced injury in other organ systems.

While the cause of NSF still remains a mystery, the investigation continues to yield important clues that are piecing together disparate clinical and histological observations into a new conceptual model of cutaneous fibrosis. The prospect that fibrosis in multiple organ systems may be united by the concept of the circulating fibrocyte is of great importance — suggesting the ideal possibility that the successes achieved in the understanding and treatment of fibrosis in other organ systems may have direct applicability to *all* organ systems.

Acknowledgment

This research was supported in part by a General Clinical Research Center grant from the National Center of Research Resources, National Institute of Health (Grant # M01-RR00125) awarded to Yale University School of Medicine.

References

1. Leboit PE. (2003) What nephrogenic fibrosing dermopathy might be. *Archiv Dermatol* **139**(7): 928–930.

2. Cowper SE, Robin HS, Steinberg SM, *et al.* (2000) Scleromyxoedema-like cutaneous disease in renal-dialysis patients. *Lancet* **356**(9234): 1000–1001.
3. Cowper SE, Su L, Robin H, Bhawan J, LeBoit PE. (2001) Nephrogenic fibrosing dermopathy. *Am J Dermatopathol* **23**(5): 383–393.
4. Mackay-Wiggan JM, Cohen DJ, Hardy MA, *et al.* (2003) Nephrogenic fibrosing dermopathy (scleromyxedema-like illness of renal disease). *J Am Acad Dermatol* **48**(1): 55–60.
5. Swartz RD, Crofford LJ, Phan SH, *et al.* (2003) Nephrogenic fibrosing dermopathy: a novel cutaneous fibrosing disorder in patients with renal failure. *Am J Med* **114**(7): 563–572.
6. Obermoser G, Emberger M, Wieser M, Zelger B. (2004) Nephrogenic fibrosing dermopathy in two patients with systemic lupus erythematosus. *Lupus* **13**(8): 609–612.
7. Cowper SE. (2005) Nephrogenic systemic fibrosis: the nosological and conceptual evolution of nephrogenic fibrosing dermopathy. *Am J Kidney Dis* **46**(4): 763–765.
8. (2002) Fibrosing skin condition among patients with renal disease — United States and Europe 1997-2002. *MMWR/Morb Mortal Wkly Rep* **51**: 25–26.
9. Cowper SE. (2006) Nephrogenic fibrosing dermopathy [NFD/NSF Website]. 2001-2006. Available at http://www.icnfdr.org. Accessed 05/08/2006.
10. Gilliet M, Cozzio A, Burg G, Nestle FO. (2005) Successful treatment of three cases of nephrogenic fibrosing dermopathy with extracorporeal photopheresis. *Br J Dermatol* **152**(3): 531–536.
11. Tan AW, Tan SH, Lian TY, Ng SK. (2004) A case of nephrogenic fibrosing dermopathy. *Ann Acad Med Singapore* **33**(4): 527–529.
12. (2006) Nephrogenic fibrosing dermopathy. *Arkh Patol* [Russian] **68**(1): 42–43.
13. Ruiz-Genao DP, Pascual-Lopez MP, Fraga S, *et al.* (2005) Osseous metaplasia in the setting of nephrogenic fibrosing dermopathy. *J Cutan Pathol* **32**(2): 172–175.
14. Cowper SE. (2003) Nephrogenic fibrosing dermopathy: the first six years. *Curr Opin Rheumatol* **15**(6): 785–790.
15. Panda S, Bandyopadhyay D, Tarafder A. (2006) Nephrogenic fibrosing dermopathy: a series in a non-Western population. *J Am Acad Dermatol* **54**(1): 155–159.

16. Cowper SE, Boyer PJ. (2006) Nephrogenic systemic fibrosis: an update. *Curr Rheumatol Rep* **8**(2): 151–157.
17. Streams BN, Liu V, Liegeois N, Moschella SM. (2003) Clinical and pathological features of nephrogenic fibrosing dermopathy. *J Am Acad Dermatol* **48**(1): 42–47.
18. Cowper SE, Bucala R. (2003) Nephrogenic fibrosing dermopathy: Suspect identified, motive unclear. *Am J Dermatopathol* **25**(4): 358.
19. Baron PW, Cantos K, Hillebrand DJ, *et al.* (2003) Nephrogenic fibrosing dermopathy after liver transplantation successfully treated with plasmapheresis. *Am J Dermatopathol* **25**(3): 204–209.
20. DeHoratius DM, Cowper SE. (2006) Nephrogenic systemic fibrosis: an emerging threat among renal patients. *Semin Dial* **19**(3): 191–194.
21. McNeill AM, Barr RJ. (2002) Scleromyxedema-like fibromucinosis in a patient undergoing hemodialysis. *Int J Dermatol* **41**(6): 364–367.
22. Hubbard V, Davenport A, Jarmulowicz M, Rustin M. (2003) Scleromyxoedema-like change in four renal dialysis patients. *Br J Dermatol* **148**(3): 563–568.
23. Kucher C, Steere J, Elenitsas R, *et al.* (2006) Nephrogenic fibrosing dermopathy/nephrogenic systemic fibrosis with diaphragmatic involvement in a patient with respiratory failure. *J Am Acad Dermatol* **54**(2 Suppl): S31–S34.
24. Cowper SE, Bucala R, Leboit PE. (2006) Nephrogenic fibrosing dermopathy/nephrogenic systemic fibrosis — setting the record straight. *Semin Arthritis Rheum* **35**(4): 208–210.
25. Gibson SE, Farver CF, Prayson RA. (2006) Multiorgan involvement in nephrogenic fibrosing dermopathy: an autopsy case and review of the literature. *Arch Pathol Lab Med* **130**(2): 209–212.
26. Ting WW, Stone MS, Madison KC, Kurtz K. (2003) Nephrogenic fibrosing dermopathy with systemic involvement. *Archiv Dermatol* **139**(7): 903–906.
27. Jimenez SA, Artlett CM, Sandorfi N, *et al.* (2004) Dialysis-associated systemic fibrosis (nephrogenic fibrosing dermopathy): study of inflammatory cells and transforming growth factor $\beta 1$ expression in affected skin. *Arthritis Rheum* **50**(8): 2660–2666.
28. Ortonne N, Lipsker D, Chantrel F, *et al.* (2004) Presence of $CD45RO^+$ $CD34^+$ cells with collagen synthesis activity in nephrogenic fibrosing dermopathy: a new pathogenic hypothesis. *Br J Dermatol* **150**(3): 1050–1052.

29. Kucher C, Xu X, Pasha T, Elenitsas R. (2005) Histopathologic comparison of nephrogenic fibrosing dermopathy and scleromyxedema. *J Cutan Pathol* **32**(7): 484–490.
30. Evenepoel P, Zeegers M, Segaert S, *et al.* (2004) Nephrogenic fibrosing dermopathy: a novel disabling disorder in patients with renal failure. *Nephrol Dial Transplant* **19**(2): 469–473.
31. Gremmels JM, Kirk GA. (2004) Two patients with abnormal skeletal muscle uptake of Tc-99m hydroxymethylene diphosphonate following liver transplant: nephrogenic fibrosing dermopathy and graft vs host disease. *Clin Nucl Med* **29**(11): 694–697.
32. Levine JM, Taylor RA, Elman LB, *et al.* (2004) Involvement of skeletal muscle in dialysis-associated systemic fibrosis (nephrogenic fibrosing dermopathy). *Muscle Nerve* **30**(5): 569–577.
33. Teder P, Noble PW. (2000) A cytokine reborn? Endothelin-1 in pulmonary inflammation and fibrosis. *Am J Respir Cell Mol Biol* **23**(1): 7–10.
34. Ottosson-Seeberger A, Ahlborg G, Hemsen A, *et al.* (1999) Hemodynamic effects of endothelin-1 and big endothelin-1 in chronic hemodialysis patients. *J Am Soc Nephrol* **10**(5): 1037–1044.
35. Abe R, Donnelly SC, Peng T, *et al.* (2001) Peripheral blood fibrocytes: differentiation pathway and migration to wound sites. *J Immunol* **166**(12): 7556–7562.
36. Hartlapp I, Abe R, Saeed RW, *et al.* (2001) Fibrocytes induce an angiogenic phenotype in cultured endothelial cells and promote angiogenesis *in vivo*. *FASEB J* **15**(12): 2215–2224.
37. Quan TE, Cowper SE, Bucala R. (2006) The role of circulating fibrocytes in Fibrosis. *Curr Rheum Reports* **8**(2): 145–150.
38. Yanagisawa M, Kurihara H, Kimura S, *et al.* (1988) A novel potent vasoconstrictor peptide produced by vascular endothelial cells. *Nature* **332**(6163): 411–415.
39. Dhaun N, Goddard J, Webb DJ. (2006) The endothelin system and its antagonism in chronic kidney disease. *J Am Soc Nephrol* **17**: 943–955.
40. Horstmeyer A, Licht C, Scherr G, *et al.* (2005) Signalling and regulation of collagen I synthesis by ET-1 and TGF-β1. *FEBS J* **272**(24): 6297–6309.
41. Farkas K, Nemcsik J, Kolossvary E, *et al.* (2005) Impairment of skin microvascular reactivity in hypertension and uraemia. *Nephrol Dial Transplant* **20**(9): 1821–1827.

42. Lundblad R, Moen O, Fosse E. (1997) Endothelin-1 and neutrophil, activation during heparin-coated cardiopulmonary bypass. *Ann Thorac Surg* **63**(5): 1361–1367.
43. Grobner T. (2006) Gadolinium — a specific trigger for the development of nephrogenic fibrosing dermopathy and nephrogenic systemic fibrosis? *Nephrol Dial Transplant* **21**(4): 1104–1108.
44. Fazeli A, Lio PA, Liu V. (2004) Nephrogenic fibrosing dermopathy: are ACE inhibitors the missing link? [Letter] *Arch Derm* **140**(11): 1401.

Chapter 12

$CD34^+$ Fibrocytes in Normal and Neoplastic Human Tissues

Peter J Barth*

Introduction

The connective tissue of virtually all human organs harbors large amounts of resident $CD34^+$ fibrocytes. The histogenesis of this cell population was not precisely defined until 1994, when Bucala and coworkers were the first to report a blood borne mesenchymal cell derived from mononuclear cells capable of tissue invasion.[1] Subsequent studies revealed that this cell — designated $CD34^+$ fibrocyte — is involved in wound healing,[1,2] acts as an antigen presenting cell[3] and secretes a multitude of cytokines.[4,5] Due to their pleiotropic functions, $CD34^+$ fibrocytes play a role in various disease processes such as systemic and localized fibroses, including scleroderma,[6] pulmonary fibrosis,[7,8] and nephrogenic fibrosing dermatopathy.[6,9] Moreover, $CD34^+$ fibrocytes are involved in stromal remodeling, precipitated by invasive carcinomas and characterized by a loss of

*Professor of Pathology, Institute of Pathology, University Hospital Giessen and Marburg GmbH, Location Marburg, Medical Faculty of Philipps-University Marburg, Baldingerstraße, 35033 Marburg, Germany.

$CD34^+$ expression paralleled by a gain of α-smooth muscle actin (α-SMA) expression in stromal cells. This results in a phenotype change from $CD34^+$ fibrocytes to α-SMA-positive myofibroblasts.[10–22] This process is highly stereotypic and may play an essential role in local tumor invasion and systemic dissemination, since a reduction of antigen presenting and matrix stabilizing $CD34^+$ fibrocytes might constitute pivotal steps in tumor cells escaping the host's immune control directed against invasive carcinoma cells.

The effort, however, to provide a complete review of the literature pertaining to $CD34^+$ fibrocytes, their morphology and function, is hampered by several methodologic problems. A clear-cut definition and distinction of fibrocytes, fibroblasts, myofibroblasts and other stromal cells is lacking and may remain somewhat arbitrary despite all scientific efforts.

In addition, the strongest data concerning $CD34^+$ fibrocytes are from animal experiments and it is arguable as to whether, and to what extent, these data apply to humans.

The tentativeness of scientists in describing stromal cells is reflected by the plethora of terms such as dendritic interstitial cells,[23] $CD34^+$ stromal cells,[15,17,21,25] adventitial fibroblastic cells,[19,24] or simply fibroblasts[26] which may denominate what we now call the $CD34^+$ fibrocyte. The cited list is by no means complete, and the diversity of terms underlines that a precise definition of stromal cells is mandatory. In the present text, the term $CD34^+$ fibrocyte will pragmatically be applied to any cell which fulfils two major criteria, the first of which is CD34 expression as detected by immunohistochemistry and the second consisting in characteristic morphologic features. This approach may implicate a wide range of studies considered to deal with $CD34^+$ fibrocytes, but as long as a precise definition is lacking, it appears appropriate.

Normal $CD34^+$ Fibrocytes — Morphology

With the advent of immunohistochemistry, CD34 positive mesenchymal cells were sporadically reported in various organs. These reports were purely descriptive in nature and in depth analyses as to the histogenesis and function of this cell type have not been undertaken. Subsequently it appeared that in humans, the connective tissue of virtually all anatomical sites such as the gastrointestinal tract,[10,12,17,23–25] lung,[15] and upper aerodigestive tract,[18] skin,[27,28] breast,[11,16,20,22,29,30] and genital tract harbors[14,31,32] great amounts of resident $CD34^+$ fibrocytes (Table 1). In serosal membranes such as the pleura and peritoneum $CD34^+$, fibrocytes form a dense submesothelial layer.[33] However, the morphology and number of $CD34^+$ fibrocytes differ slightly depending on the anatomical site (Fig. 1). Mostly,

Table 1 Studies concerning the Occurrence of $CD34^+$ Fibrocytes in Normal Tissues and the Carcinoma-associated Stroma

Anatomical Site	References	Loss of $CD34^+$ Fibrocytes in Carcinomas
Skin	9, 27, 28	+
Breast	11, 16, 20, 22, 29, 30	IDC: +, ILC: ±
Resiratory tract		
Larynx	18	+
Lung	7, 8, 15	+
Heart	34	n.d.
Thyroid	36	n.d.
Pleura		
Gastrointestinal tract		
Salivary glands	23, 35	+
Oral cavity	18	+
Pharynx	18	+
Stomach	13, 19, 24, 25	IT: +, DT: ±
Gall bladder		
Pancreas	12, 17	+, NET: +
Small intestine		
Colon	10, 21	AC: +, NET: +
Peritoneum	33	+
Genital tract		
Cervix	14	+
Fallopian tube	31	n.d.
Testis	32	n.d.
Eye	41	n.d.

IDC: invasive ductal carcinoma; ILC: invasive lobular carcinoma; IT: intestinal type carcinoma; DT: diffuse carcinoma; AC: adenocarcinoma; NET: neuroendocrine tumor; +: consistent lack of $CD34^+$ fibrocytes; ±: partially preserved $CD34^+$ fibrocytes; n.d.: not done.

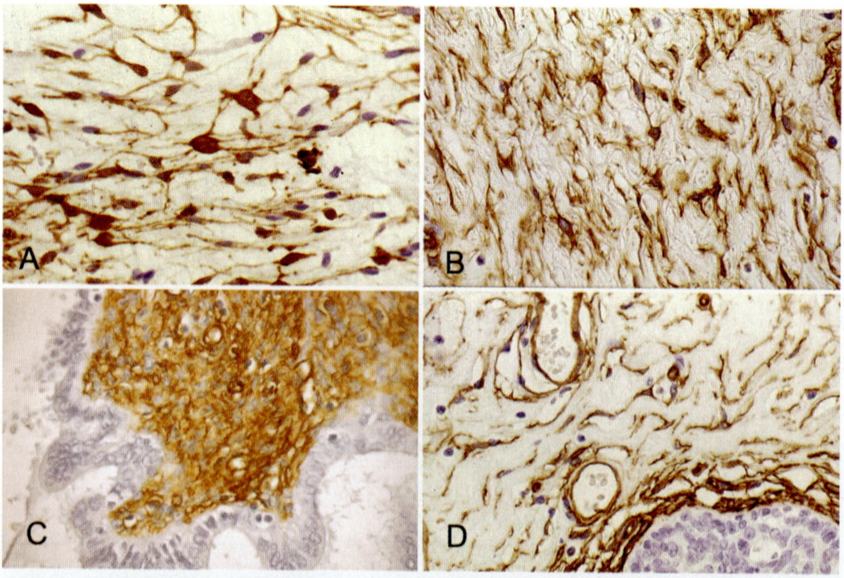

Fig. 1. CD34$^+$ fibrocytes show multiple slender and communicating cytoplasmic processes in the loose connective tissue of the vocal cord (**A**). The collagen fiber-rich cervical stroma shows CD34$^+$ fibrocytes arranged in parallel (**B**). Focally, the cells form a dense network rendering the identification of individual cells impossible (**C**). Mammary acini are encircled by densely packed CD34$^+$ fibrocytes; the extralobular stroma harbors few dispersed CD34$^+$ fibrocytes (**D**). (**A**)–(**D**): CD34 immunostaining.

CD34$^+$ fibrocytes show a small inconspicious cytoplasma forming long slender bi- or multipolar cytoplasmic processes (Fig. 1A). The nucleus is small, slightly elongated and shows a lacy or finely granular chromatin structure; the cytoplasm is scarce.[11,12,14,16,18,20] The length of the individual CD34$^+$ fibrocytes ranges between 60 and 120 μm (arithmetic mean: 90 μm).

In collagen fiber-rich connective tissue such as the cervix uteri[14] or the fibrosa of heart valves,[34] CD34$^+$ fibrocytes are arranged in parallel between adjacent broad bands of collagen (Fig. 1B). In the lamina propria of hollow organs such as the larynx,[18] pharynx,[18] oral cavity,[18] fallopian tube,[31] vagina, cervix and urinary bladder, few evenly dispersed multipolar CD34$^+$ fibrocytes are found in the loose connective tissue adjacent to the covering epithelium, whereas in the deeper and mostly more collagen fiber-rich areas, CD34$^+$ fibrocytes appear to be more closely packed (Fig. 1C). The same distribution pattern was observed in the submucosa of the stomach and colon, where the highest numbers of CD34$^+$ fibrocytes are found around arteries and to a lesser extent veins, prompting the descriptive term "adventitial fibroblastic cells" by some investigators.[19,24] Fascicles of smooth muscle making up the tunica muscularis are surrounded by CD34$^+$ fibrocytes. In contrast, the mucosa of the stomach and colon are devoid of CD34$^+$ fibrocytes.[10,21,25]

In the breast, the intralobular stroma harbors large numbers of densely packed CD34$^+$ fibrocytes encircling ducts and acini (Fig. 1D). The extralobular stroma, in contrast, shows few CD34$^+$ fibrocytes with a perivascular predominance as described above.[11,16,20] A similar distribution of CD34$^+$ fibrocytes was found in the pancreas.[12] In general, glandular organs with a lobular architecture reveal the highest amounts of CD34$^+$ fibrocytes in the intralobular stroma, where they surround acinar and ductal structures. The extralobular stroma, which is mostly less cell rich, shows few CD34$^+$ fibrocytes, although most intra- and extralobular stromal cells are CD34 positive.

Immunohistochemically, normal CD34$^+$ fibrocytes stain for vimentin,[1] whereas no positivity for α-SMA or desmin was detected in most studies.[11,12,14,16,18,20] However, few studies reporting a coexpression of α-SMA and CD34 in stromal cells;[21,32] this issue remains controversial and may further pose a semantic problem since it is unclear whether α-SMA positive stromal cells are CD34$^+$ fibrocytes according to the definition of this cell type. In heart valves, CD34$^+$ fibrocytes also stain for S100 protein,[34] an expression pattern which is peculiar to but not specific for this anatomical site since it is also observed in lipomatous tumors.[26] CD117 expression is absent from CD34$^+$ fibrocytes, distinguishing them from Cajal's pacemaker cells of the gastrointestinal tract[18] (Table 2).

On the ultrastructural level, Yamazaki and Eyden using immunoelectron-microscopy demonstrated CD34$^+$ fibrocytes, referred to as fibroblasts by them, in the breast,[29] fallopian tube,[31] submandibular gland,[35] and thyroid,[36] where they formed a peculiar reticular network composed of slender communicating cytoplasmic processes

Table 2 Morphologic Features and Immunohistochemical Profile (IMH) of Carcinoma-associated Stromal Remodeling

	Normal	Carcinoma-associated
Morphology	Fibrocyte	Myofibroblast
IMH	Antigen (function)	Antigen (function)
	Vimentin (intermediate filament)	Vimentin
	CD 34 (adhesion, L-selectin ligand)	α-SMA (stromal contration)
	bcl-2* (anti-apoptotic)	SPARC (stromal de-adhesion)
		CD117 (growth factor receptor)**

*: pertinent data restricted to the breast; **: ligand: stem cell factor.

described as "CD34 positive reticular network."[31,32] Cytoplasmic processes of neighboring CD34$^+$ fibrocytes were attached by gap junctions, and close membrane appositions were found between mononuclear cells and the CD34 positive reticular network, leading to the assumption that the CD34 positive reticular network might play a role in host immunosurveillance.[32,35,36]

CD34$^+$ Fibrocytes in the Carcinoma-associated Stroma

Early systematic studies describing immunophenotypic alterations in stromal CD34 expression were performed in cutaneous lesions.[27,28] In these studies, CD34$^+$ fibrocytes were given different descriptive terms, leading to some confusion as to what cell type was precisely meant and rendering the comparison of various studies difficult. However, in general, the stroma adjacent to basal cell carcinomas was characterized by a loss of CD34$^+$ fibrocytes, whereas that associated with benign lesions showed a preserved population of CD34$^+$ fibrocytes.[27,28] Later studies revealed this phenomenon not to be restricted to the skin but to occur at virtually any anatomic location investigated until now.

Irrespective of the primary site of the tumor and histologic type of the carcinoma, the stroma associated with invasive carcinomas stereotypically reveals a loss of CD34$^+$ fibrocytes.[10–12,14–20] Most carcinomas investigated as to this topic were adeno-[10–12,15–17,19,20] or squamous cell carcinomas,[14,18] and in the more rare endocrine carcinomas, a loss of CD34$^+$ fibrocytes was described.[12,21]

Histologically, the border between tumor-free tissue and invasive carcinomas is mostly characterized by an abrupt loss of CD34$^+$ fibrocytes[11,12,14,16,18,20] (Fig. 2A). The stromal cells undergo a transition from the CD34$^+$ α-SMA$^-$ phenotype, towards a CD34$^-$ ASMA$^+$ phenotype, accompanied by morphologic alterations characterized by a loss of the slender dendrite-like cytoplasmic processes and adoption of a plump spindle cell appearance (Fig. 2B). The stroma adjacent to *in situ* carcinomas of the breast (DCIS)[11] and cervix

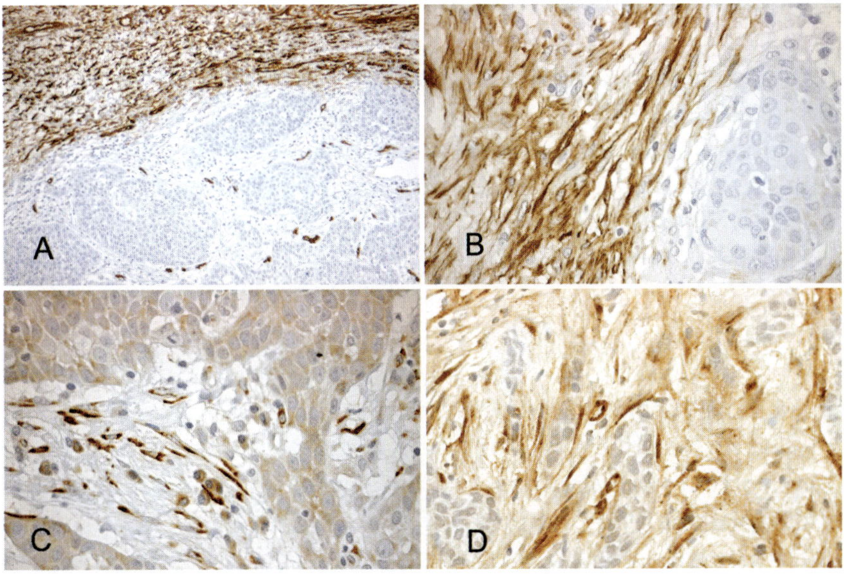

Fig. 2. The border between tumor-free stroma and carcinoma-associated stroma is characterized by an abrupt loss of CD34$^+$ fibrocytes ((**A**), squamous cell carcinoma of the oral cavity). The loss of stromal CD34 expression is paralleled by a gain of α-SMA expression ((**B**), invasive ductal carcinoma of the breast). Strong cytoplasmic immunostaining for CD117 in a squamous cell carcinoma of the upper aerodigestive tract (**C**). SPARC is constantly found in the stroma of invasive ductal carcinomas of the breast (**D**).

(CIN III)[14] shows similar alterations, suggesting a diffusible factor possibly mediating this phenomenon. In contrast, endothelial cells of vessels located within the tumor show unaltered CD34 reactivity, a finding which excludes artefacts or simple enzymatic shedding as an explanation of the observed loss of CD34 expression.[12] Moreover, the reported findings appear to be independent of the chosen antibody clone used to detect the CD34 antigen. Until recently, it was unclear as to whether a population of stromal cells with dual expression of CD34 and α-SMA exists which might constitute a hybrid or transition form between $CD34^+$ fibrocytes and α-SMA positive myofibroblasts. When investigating stromal alterations at the border between benign and malignant salivary gland tumors, Soma and coworkers[23] found two strictly distinct populations of $CD34^+$ "dendritic interstitial cells," i.e. fibrocytes and α-SMA reactive myofibroblasts. Apparently no cells with co-expression of the two markers were found, strengthening the notion that expression of CD34 and α-SMA are mutually exclusive and suggesting that a cell type with dual expression does not exist.

Reports of a partially preserved population of $CD34^+$ fibrocytes within carcinoma-associated stroma are rare. In diffuse-type gastric adenocarcinoma, the stromal $CD34^+$ fibrocyte population appears to be preserved[24,25] and our own preliminary data suggest that this might be true for a subpopulation of lobular carcinomas of the breast as well. Interestingly, both these entities are characterized by the absence of E-cadherin which may be responsible for their disseminated growth pattern and tissue invasion.

The loss of $CD34^+$ fibrocytes is further accompanied by additional phenotypic alterations of the stromal cells, such as a gain of cytoplasmic CD117 expression in a subpopulation of squamous cell carcinomas of the upper aerodigestive tract[18] (Fig. 2C, Table. 2). Whereas the latter finding appears to be restricted to a relatively small number of entities, secreted protein acid rich in cysteine (SPARC), also referred to as ostenectin or BM-40, is found in the stromal cells of invasive ductal carcinomas of the breast[20] (Fig. 2D, Table 2) and, as recent studies have shown, virtually all invasive carcinomas.

Pathogenesis of CD34+ Fibrocyte Loss

The aforementioned loss of CD34+ fibrocytes occurs in most carcinomas explored until now, irrespective of their being invasive or *in situ*, indicating that a loss of the stromal CD34+ fibrocyte network is an important prerequisite of local tumor invasion and final distant systemic spread. As CD34+ fibrocytes are capable of collagen synthesis,[1,2,4,5] they play a major role in matrix repair opposing matrix-metalloproteinases (MMPs) secreted by invasive carcinoma cells. Second, since CD34+ fibrocytes act as antigen presenting cells,[3] the reduction of this cell population may constitute a pivotal factor in the tumor's escape from host immunosurveillance. Neuroendocrine pulmonary tumor cells, for example, have been reported to be capable of inducing apoptosis in dendritic stromal cells closely related to the CD34+ fibrocytes by a not yet precisely defined soluble factor.[37] The mechanisms, however, precipitating the loss of CD34+ fibrocyte in the tumor-associated stroma, are far from being completely understood. There are, nevertheless, two factors, TGF-β and SPARC[38–40] that have been shown to play an important role in stromal remodeling precipitated by invasive carcinomas.

In cell and tissue culture studies, TGF-β upregulates α-SMA[2] and downregulates CD34 expression.[33,41] Assuming that this is also the fact *in vivo*, the immunophenotypic alterations occurring in the carcinoma-associated stroma might putatively be caused by TGF-β. The number of mast cells, which are a major source of TGF-β, is significantly increased in the stroma of squamous cell carcinomas of the upper aerodigestive tract[18] and cutaneous basal cell carcinomas,[27] and the expression of the TGF-β-receptor and SPARC were shown to be increased in the stroma of the squamous cell carcinomas[18] (Fig. 2D). SPARC, a matricellular protein, the expression of which is up-regulated by TGF-β[42,43] has been shown, by means of cDNA microarrays, to be a member of the breast cancer "invasion-specific" cluster.[38] In the context of stromal remodeling associated with invasive carcinoma, the most important function of SPARC is to precipitate de-adhesion of the carcinoma-associated stroma,[44] which is defined by an intermediate state of adhesion caused by restructuring of focal adhesions.[45] De-adhesion of stromal cells also precipitates alterations of the cytoskeleton, which are in part characterized by an increased synthesis of stress fibers. This phenomenon is reflected by the abundance of α-SMA positive intracytoplasmic fibers,[45] which may be responsible for the alterations of the stromal cell shape and the adoption of a plump, spindle cell appearance. SPARC may account for these properties and is believed to constitute a pivotal factor in tumor cell invasion and tumor cell migration.[46] SPARC and TGF-β expression are closely related and mutually influence each other.[47] It is likely that both TGF-β and SPARC are important regulators of carcinoma-associated stromal remodeling (Fig. 3).

The carcinoma-associated loss of CD34$^+$ fibrocytes stereotypically appears in virtually all invasive carcinomas meanwhile

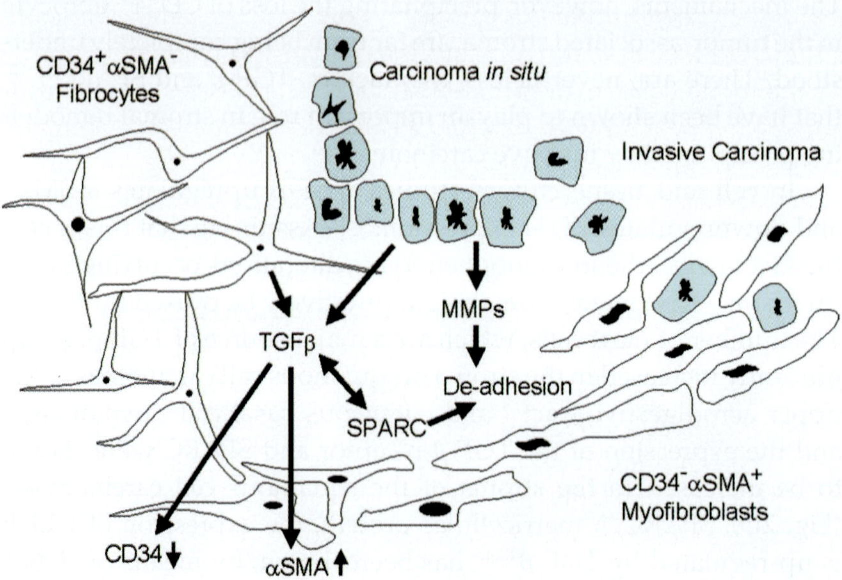

Fig. 3. TGF-β and SPARC act together in carcinoma-associated stromal remodeling, morphologically supporting a transition from CD34$^+$ α-SMA$^-$ fibrocytes to CD34$^-$ α-SMA$^+$ myofibroblasts. Downregulation of CD34 and upregulation of α-SMA are mainly due to TGF-β, whereas SPARC and MMPs mediate stromal de-adhesion, which precedes tumor cell invasion.

investigated. One may therefore conclude that it constitutes an important prerequisite of local tumor invasion and systemic spread, since CD34$^+$ fibrocytes are involved in matrix-synthesis. Since a preserved network of CD34$^+$ fibrocytes supports stromal integrity, local tumor invasion and distant spread might be impossible unless the function of CD34$^+$ fibrocytes is abrogated by induction of a CD34 negative phenotyope and subsequent stromal de-adhesion.

Diagnostic Significance of CD34$^+$ Fibrocytes

Several studies have addressed the diagnostic value of the immunohistochemical detection of CD34$^+$ fibrocytes. Complex sclerosing lesions and radial scars of the breast regularly pose significant diagnostic challenges, and the combined detection of CD34$^+$ fibrocytes and αSMA$^+$myofibroblasts has been shown to constitute a valuable adjunctive tool in distinguishing these from tubular carcinomas.[16] As mentioned above, the differential diagnosis between benign skin appendage tumors with basaloid features and cutaneous basal cell carcinoma can be improved by detection of CD34$^+$ fibrocytes in the stroma adjacent to benign lesions, whereas the desmoplastic stroma characteristic of basal cell carcinomas lacks this cell type.[27,28] Another diagnostic dilemma likely solved by the detection of CD34$^+$ fibrocytes is the distinction of chronic pancreatitis and ductal pancreatic carcinoma. In pancreatic ductal carcinomas, a loss of CD34$^+$ fibrocytes occurs, whereas this cell population is increased in chronic pancreatitis, indicating that pancreatic fibrosis might be due in part to the capacity of CD34$^+$ fibrocytes to synthesize collagen.[12]

Peritoneal implants from borderline ovarian tumors can be well distinguished from peritoneal carcinomatosis derived from frankly malignant papillary serous ovarian carcinoma.[48] However, in considering the detection of CD34$^+$ fibrocytes as a diagnostic tool, it has to be kept in mind that granulation tissue occurring at the site of a previous core needle biopsy in the breast immunophenotypically resembles tumor-associated stroma in that a loss of CD34$^+$ fibrocytes can be found together with newly occurring α-SMA$^+$ myofibroblasts.[30] The same phenomenon occurs in xanthogranulomatous cholecystitis.[49] Therefore, a loss of CD34$^+$ fibrocytes should not be used as the sole diagnostic criterion of malignancy. In contrast, the presence of CD34$^+$ fibrocytes excludes malignancy with a high degree of probability since only few malignant lesions (e.g. lobular carcinoma of the breast and diffuse adenocarcinoma of the stomach) have been described in which a normal population of CD34$^+$ fibrocytes appears to be at least in part conserved.

Tumors Histogenetically Linked to CD34$^+$ Fibrocytes

The assumption that a certain neoplastic lesion might be histogenetically related to CD34$^+$ fibrocytes is mainly based on two grounds: first, CD34 expression as detected by immunohistochemistry, and second, histologic features displaying at least a focal spindle cell population morphologically resembling CD34$^+$ fibrocytes. Most lesions summarized in this group, except for solitary fibrous tumors (SFTs) which show a wide anatomical distribution, are located in the skin, breast and subcutaneous tissue. Taken together, an astonishingly high number of mostly fibrous and lipomatous soft tissue tumors have been claimed to be derived from CD34$^+$ fibrocytes.

Solitary Fibrous Tumor (SFT)

Although relatively rare compared to some other tumors discussed in this section, solitary fibrous tumors (SFTs) appear to be the most characteristic and important member of the group of mesenchymal tumors putatively derived from CD34$^+$ fibrocytes. Many features of solitary fibrous tumors (SFTs) suggest these benign lesions to be histogenetically derived from, or at least closely related to, CD34$^+$ fibrocytes. Although these tumors have primarily been described as pleural lesions, they may develop in a wide range of anatomical sites.[50,51] Histologically, SFTs are composed of small spindled mesenchymal cells capable of abundant collagen synthesis resembling CD34$^+$ fibrocytes. Immunohistochemically, SFTs display intense staining for CD34 and CD99, whereas α-SMA, cytokeratins and S100 are absent.[51]

Dermatofibrosarcoma Protuberans (DFSP)

As in SFTs the close relationship of DFSP tumors with $CD34^+$ fibrocytes is unequivocal. Dermatofibrosarcoma protuberans (DFSP) is a tumor involving the skin and subcutaneous tissues that is composed of a monomorphic, uniform spindle cell population closely resembling typical $CD34^+$ fibrocytes. Immunohistochemically, the tumor shows strong CD34 positivity.[52] Since fibrosarcomatous or malignant fibrous histiocytomatous transformation of DFSP has rarely, but conclusively been described, it has been argued that fibrosarcoma (FS) and malignant fibrous histiocytoma (MFH) might be related to $CD34^+$ fibrocytes.[53,54] However, since most MFHs and FSs develop *de novo* and lack CD34 immunoreactivity, this presumption appears doubtful and the histogenesis of MFHs and FSs should be regarded to be a matter for further investigation.

Stromal Tumors of the Breast

In the breast, the stroma of fibroadenomas and phyllodes tumors comprises a $CD34^+$ cell population closely resembling $CD34^+$ fibrocytes of the surrounding mammary tissue.[11,30,55] Benign spindle stromal tumors (BSSTs) of the breast such as spindle cell lipoma-like tumor, solitary fibrous tumor and myofibroblastoma, have also been postulated as deriving from $CD34^+$ fibrocytes.[56] This assumption has been underlined by the finding that at least in the breast, $CD34^+$ fibrocytes show co-expression of bcl-2, suggesting a putative progenitor cell function of $CD34^+$ fibrocytes.[55,56]

Lipomatous Tumors

It has been claimed that lipomatous tumors such as spindle cell, pleomorphic, myxoid, and atypical lipomas as well as dedifferentiated liposarcomas, owing to their CD34 immunopositivity, are derived from CD34 positive mesenchymal, cells.[26,56,57] This is further strengthened by the finding that both mature adipocytes and preadipocytes are CD34 positive.[58,59]

Miscellaneous Tumors

Varying populations of CD34 reactive spindle cells occur in angiomyofibroblastoma and aggressive angiomyxoma, which are predominantly located in the vagina, vulva and perineum.[60] However, in contrast to the aforementioned lipomatous tumors, stromal tumors of the breast and SFTs, the histogenetic relation to CD34$^+$ fibrocytes appears not to be sufficiently established at present.

Gastrointestinal Stromal Tumors

Gastrointestinal stromal tumors (GISTs) arise from gastrointestinal interstitial cells, also termed "Cajal cells" according to their first description by Ramon Cajal (1852–1934). These cells exert pacemaker functions and regulate gastrointestinal peristalsis. Morphologically, Cajal cells resemble CD34$^+$ fibrocytes but have even more slender and dendrite-like cytoplasmic projections. They display CD34 expression, but in addition, show CD117 positivity which clearly distinguishes them from CD34$^+$ fibrocytes. Accordingly, GISTs mostly express CD34 and CD117. Moreover, about 30% of GISTs exhibit αSMA positivity, indicating that Cayal cells might in addition to their pacemaker function putatively act as progenitors of smooth muscle. Nevertheless, GISTs are clearly distinct from leiomyomas and leiomyosarcomas since these tumors lack expression of CD34 and CD117.[61]

Concluding Remarks and Future Perspectives

In addition to their role in matrix repair, antigen presentation and carcinoma-associated stromal remodelling, CD34$^+$ fibrocytes appear to serve a progenitor function for mesenchymal tissues, an aspect underlined by the multitude of predominantly benign mesenchymal tumors related to CD34$^+$ fibrocytes. However, the most intriguing aspect of CD34$^+$ fibrocytes is their strong expression of CD34, a sialomucin whose functions are up to now not completely understood. In endothelial cells, there is evidence that CD34 serves as an

L-selectin ligand mediating leucocyte tissue immigration and homing to lymphatic organs.[62] The function of CD34 in the stroma, in contrast, is completely enigmatic and a matter of speculation. Since CD34 mediates cell adhesion,[63] it might be important in stabilizing the stromal CD34$^+$ reticular network and thus counteracting tumor cell migration and tissue invasion. Secondly, the CD34$^+$ reticular network might guide leukocyte migration via direct mechanical interaction[62] and chemokine secretion.[4,5] The precise knowledge as to the function and morphology of CD34$^+$ fibrocytes will therefore significantly contribute to the understanding of chronic inflammatory diseases, invasiveness and systemic spread of carcinomas, and the pathogenesis of soft tissue tumors. Furthermore, CD34$^+$ fibrocytes may provide a diagnostic tool, the evaluation of which requires further scrutiny.

References

1. Bucala R, Spiegel LA, Chesney J, *et al.* (1994) Circulating fibrocytes define a new leukocyte subpopulation that mediates tissue repair. *Mol Med* **1**: 71–81.
2. Abe R, Donnelly SC, Peng T, *et al.* (2001) Peripheral blood fibrocytes: differentiation pathway and migration to wound sites. *J Immunol* **166**: 7556–7562.
3. Chesney J, Bacher M, Bender A, Bucala R. (1997) The peripheral blood fibrocyte is a potent antigen-presenting cell capable of priming naive T cells *in situ*. *Proc Natl Acad Sci USA* **94**: 6307–6312.
4. Chesney J, Metz C, Stavitsky AB, *et al.* (1998) Regulated production of type I collagen and inflammatory cytokines by peripheral blood fibrocytes. *J Immunol* **160**: 419–425.
5. Hartlapp I, Abe R, Saeed RW, *et al.* (2001) Fibrocytes induce an angiogenic phenotype in cultured endothelial cells and promote angiogenesis *in vivo*. *FASEB J* **15**: 2215–2224.
6. Quan TE, Cowper S, Wu SP, *et al.* (2004) Circulating fibrocytes: collagen-secreting cells of the peripheral blood. *Int J Biochem Cell Biol* **36**: 598–606.
7. Schmidt M, Sun G, Stacey MA, *et al.* (2003) Identification of circulating fibrocytes as precursors of bronchial myofibroblasts in asthma. *J Immunol* **171**: 380–389.

8. Phillips RJ, Burdick MD, Hong K, *et al.* (2004) Circulating fibrocytes traffic to the lungs in response to CXCL12 and mediate fibrosis. *J Clin Invest* **114**: 438–446.
9. Daram SR, Cortese CM, Bastani B. (2005) Nephrogenic fibrosing dermopathy/nephrogenic systemic fibrosis: report of a new case with literature review. *Am J Kidney Dis* **46**(4): 754–759.
10. Nakayama H, Enzan H, Miyazaki E, *et al.* (2000) Differential expression of CD34 in normal colorectal tissue, peritumoral inflammatory tissue, and tumour stroma. *J Clin Pathol* **53**: 626–629.
11. Barth PJ, Ebrahimsade S, Ramaswamy A, Moll R. (2002) $CD34^+$ fibrocytes in invasive ductal carcinoma, ductal carcinoma in situ, and benign breast lesions. *Virchows Arch* **440**: 298–303.
12. Barth PJ, Ebrahimsade S, Hellinger A, *et al.* (2002) $CD34^+$ fibrocytes in neoplastic and inflammatory pancreatic lesions. *Virchows Arch* **440**: 128–133.
13. Nakayama H, Enzan H, Miyazaki E, Toi M. (2002) Alpha smooth muscle actin positive stromal cells in gastric carcinoma. *J Clin Pathol* **55**: 741–744.
14. Barth PJ, Ramaswamy A, Moll R. (2002) $CD34^+$ fibrocytes in normal cervical stroma, cervical intraepithelial neoplasia III, and invasive squamous cell carcinoma of the cervix uteri. *Virchows Arch* **441**: 564–568.
15. Nakayama H, Enzan H, Yamamoto M, *et al.* (2003) CD34-positive stromal cells in primary lung carcinomas. *Oncol Rep* **10**: 1313–1316.
16. Ramaswamy A, Moll R, Barth PJ. (2003) $CD34^+$ fibrocytes in tubular carcinomas and radial scars of the breast. *Virchows Arch* **443**: 536–540.
17. Kuroda N, Toi M, Nakayama H, *et al.* (2004) The distribution and role of myofibroblasts and CD34-positive stromal cells in normal pancreas and various pancreatic lesions. *Histol Histopathol* **19**: 59–67.
18. Barth PJ, Schenck zu Schweinsberg T, *et al.* (2004) $CD34^+$ fibrocytes, alpha-smooth muscle antigen-positive myofibroblasts, and CD117 expression in the stroma of invasive squamous cell carcinomas of the oral cavity, pharynx, and larynx. *Virchows Arch* **444**: 231–234.
19. Nakayama H, Enzan H, Miyazaki E, *et al.* (2004) Lack of vascular adventitial fibroblastic cells in tumour stroma of intestinal-type and solid-type gastric carcinomas. *J Clin Pathol* **57**: 183–185.

20. Barth PJ, Moll R, Ramaswamy A. (2005) Stromal remodeling and SPARC (secreted protein acid rich in cysteine) expression in invasive ductal carcinomas of the breast. *Virchows Arch* **446**: 532–536.
21. Kuroda N, Nakayama H, Miyazaki E, *et al.* (2005) The distribution of CD34-positive stromal cells and myofibroblasts in colorectal carcinoid tumors. *Histol Histopathol* **20**: 27–33.
22. Yazhou C, Wenlv S, Weidong Z, Licun W. (2004) Clinicopathological significance of stromal myofibroblasts in invasive ductal carcinoma of the breast. *Tumour Biol* **25**: 290–295.
23. Soma L, LiVolsi VA, Baloch ZW. (2001) Dendritic interstitial and myofibroblastic cells at the border of salivary gland tumors. *Arch Pathol Lab Med* **125**: 232–236.
24. Nakayama H, Enzan H, Miyazaki E, *et al.* (2004) Presence of vascular adventitial fibroblastic cells in diffuse-type gastric carcinomas. *J Clin Pathol* **57**: 970–972.
25. Nakayama H, Enzan H, Miyazaki E, *et al.* (2001) CD34 positive stromal cells in gastric adenocarcinomas. *J Clin Pathol* **54**: 846–848.
26. Silverman JS, Tamsen A. (1997) Fibrohistiocytic differentiation in subcutaneous fatty tumors. Study of spindlecell, pleomorphic, myxoid, and atypical lipoma and dedifferentiated liposarcoma cases composed in part of $CD34^+$ fibroblasts and $FXIIIa^+$ histiocytes. *J Cutan Pathol* **24**: 484–493.
27. Humphreys TR, Monteiro MR, Murphy GF. (2000) Mast cells and dendritic cells in basal cell carcinoma stroma. *Dermatol Surg* **26**: 200–203.
28. Kirchmann TT, Prieto VG, Smoller BR. (1994) CD34 staining pattern distinguishes basal cell carcinoma from trichoepithelioma. *Arch Dermatol* **130**: 589–592.
29. Yamazaki K, Eyden BP. (1995) Ultrastructural and immunohistochemical observations on intralobular fibroblasts of human breast, with observations on the CD34 antigen. *J Submicrosc Cytol Pathol* **27**: 309–323.
30. Chauhan H, Abraham A, Phillips JR, *et al.* (2003) There is more than one kind of myofibroblast: analysis of CD34 expression in benign, in situ, and invasive breast lesions. *J Clin Pathol* **56**: 271–276.
31. Yamazaki K, Eyden BP. (1996) Ultrastructural and immunohistochemical studies of stromal cells in lamina propria of human fallopian tube ampullar mucosa: the recognition of "CD34 positive reticular

network" and its putative function in immune surveillance. *J Submicrosc Cytol Pathol* **28**: 325–337.
32. Kuroda N, Nakayama H, Miyazaki E, *et al.* (2004) Distribution and role of CD34-positive stromal cells and myofibroblasts in human normal testicular stroma. *Histol Histopathol* **19**: 743–751.
33. Jimenez-Heffernan JA, Aguilera A, Aroeira LS, *et al.* (2004) Immunohistochemical characterization of fibroblast subpopulations in normal peritoneal tissue and in peritoneal dialysis-induced fibrosis. *Virchows Arch* **444**: 247–256.
34. Barth PJ, Koster H, Moosdorf R. (2005) CD34^{+} fibrocytes in normal mitral valves and myxomatous mitral valve degeneration. *Pathol Res Pract* **201**(4): 301-304.
35. Yamazaki K, Eyden BP. (1996) Ultrastructural and immunohistochemical studies of intralobular fibroblasts in human submandibular gland: the recognition of a "CD34 positive reticular network" connected by gap junctions. *J Submicrosc Cytol Pathol* **28**: 471–483.
36. Yamazaki K, Eyden BP. (1997) Interfollicular fibroblasts in the human thyroid gland: recognition of a CD34 positive stromal cell network communicated by gap junctions and termineated by autonomic nerve endings. *J Submicrosc Cytol Pathol* **29**: 461–467.
37. Katsenelson NS, Shurin GV, Bykovskaia SN, *et al.* (2001) Human small cell lung carcinoma and carcinoid tumor regulate dendritic cell maturation and function. *Mod Pathol* **14**: 40–45.
38. Iacobuzio-Donahue CA, Argani P, Hempen PM, *et al.* (2002) The desmoplastic response to infiltrating breast carcinoma: gene expression at the site of primary invasion and implications for comparisons between tumor types. *Cancer Res* **62**: 5351–5357.
39. Porter PL, Sage EH, Lane TF, *et al.* (1995) Distribution of SPARC in normal and neoplastic human tissue. *J Histochem Cytochem* **43**: 791–800.
40. Ryu B, Jones J, Hollingsworth MA, *et al.* (2001) Invasion-specific genes in malignancy: serial analysis of gene expression comparisons of primary and passaged cancers. *Cancer Res* **61**: 1833–1838.
41. Espana EM, Kawakita T, Liu CY, Tseng SC. (2004) CD-34 expression by cultured human keratocytes is downregulated during myofibroblast differentiation induced by TGF-beta1. *Invest Ophthalmol Vis Sci* **45**: 2985–2991.
42. Reed MJ, Vernon RB, Abrass IB, Sage EH. (1994) TGF-beta 1 induces the expression of type I collagen and SPARC, and enhances contraction of

collagen gels, by fibroblasts from young and aged donors. *J Cell Physiol* **158**: 169–179.
43. Abe K, Hibino T, Mishima H, Shimomura Y. (2004) The cytokine regulation of SPARC production by rabbit corneal epithelial cells and fibroblasts in vitro. *Cornea* **23**: 172–179.
44. Bradshaw AD, Sage EH. (2001) SPARC, a matricellular protein that functions in cellular differentiation and tissue response to injury. *J Clin Invest* **107**: 1049–1054.
45. Murphy-Ullrich JE. (2001) The de-adhesive activity of matricellular proteins: is intermediate cell adhesion an adaptive state? *J Clin Invest* **107**: 785–790.
46. Framson PE, Sage EH. (2004) SPARC and tumor growth: where the seed meets the soil. *J Cell Biochem* **92**: 679–690.
47. Schiemann BJ, Neil JR, Schiemann WP. (2003) SPARC inhibits epithelial cell proliferation in part through stimulation of the transforming growth factor-β-signaling system. *Mol Biol Cell* **14**: 3977–3988.
48. Lee ES, Leong AS, Kim YS, *et al.* (2006) Calretinin, CD34, and alpha-smooth muscle actin in the identification of peritoneal invasive implants of serous borderline tumors of the ovary. *Mod Pathol* **19**: 364–372.
49. Kuroda N, Guo L, Miyazaki E, *et al.* (2005) The appearance of myofibroblasts and the disappearance of CD34-positive stromal cells in the area adjacent to xanthogranulomatous foci of chronic cholecystitis. *Histol Histopathol* **20**: 127–133.
50. Goodlad JR, Fletcher CD. (1991) Solitary fibrous tumour arising at unusual sites: analysis of a series. *Histopathology* **19**: 515–522.
51. Dotto JE, Ahrens W, Lesnik DJ, *et al.* (2006) Solitary fibrous tumor of the larynx: a case report and review of the literature. *Arch Pathol Lab Med* **30**: 213–216.
52. Haycox CL, Odland PB, Olbricht SM, Piepkorn M. (1997) Immunohistochemical characterization of dermatofibrosarcoma protuberans with practical applications for diagnosis and treatment. *J Am Acad Dermatol* **37**: 438–444.
53. Szollosi Z, Nemes Z. (2005) Transformed dermatofibrosarcoma protuberans: a clinicopathological study of eight cases. *J Clin Pathol* **58**: 751–756.
54. Silverman JS, Tamsen A. (1998) High grade malignant fibrous histiocytomas have bimodal cycling populations of factor XIIIa$^+$ dendrophages

and dedifferentiated mesenchymal cells possibly derived from CD34^{+} fibroblasts. *Cell Vis* **5**: 73–76.

55. Moore T, Lee AH. (2001) Expression of CD34 and bcl-2 in phyllodes tumours, fibroadenomas and spindle cell lesions of the breast. *Histopathology* **38**: 62–67.
56. Magro G, Bisceglia M, Michal M, Eusebi V. (2002) Spindle cell lipoma-like tumor, solitary fibrous tumor and myofibroblastoma of the breast: a clinico-pathological analysis of 13 cases in favor of a unifying histogenetic concept. *Virchows Arch* **440**: 249–260.
57. Suster S, Fisher C. (1997) Immunoreactivity for the human hematopoietic progenitor cell antigen (CD34) in lipomatous tumors. *Am J Surg Pathol* **21**: 195–200.
58. Festy F, Hoareau L, Bes-Houtmann S, *et al.* (2005) Surface protein expression between human adipose tissue-derived stromal cells and mature adipocytes. *Histochem Cell Biol* **124**: 113–121.
59. Sengenes C, Lolmede K, Zakaroff-Girard A, *et al.* (2005) Preadipocytes in the human subcutaneous adipose tissue display distinct features from the adult mesenchymal and hematopoietic stem cells. *J Cell Physiol* **205**: 114–122.
60. Silverman JS, Albukerk J, Tamsen A. (1997) Comparison of angiomyofibroblastoma and aggressive angiomyxoma in both sexes: four cases composed of bimodal CD34 and factor XIIIa positive dendritic cell subsets. *Pathol Res Pract* **193**: 673–682.
61. Riddell RH, Petras RE, Williams GT, Sobin LH. (2003) Mesenchymal tumors. In: J Rosai (ed.). *Atlas of Tumor Pathology*, 3rd series. Fascicle 32. Tumors of the intestines, pp: 325–394. Armed Forces Institute of Pathology, Washington, DC.
62. Khan AI, Landis RC, Malhotra R. (2003) L-Selectin ligands in lymphoid tissues and models of inflammation. *Inflammation* **27**: 265–280.
63. Majdic O, Stockl J, Pickl WF, *et al.* (1994) Signaling and induction of enhanced cytoadhesiveness via the hematopoietic progenitor cell surface molecule CD34. *Blood* **83**: 1226–1234.

INDEX

E

F

G

H

I

L

M

N

P

R

S